UNDERSTANDING
Earth's Deep Past

Lessons for Our Climate Future

Committee on the
Importance of Deep-Time Geologic Records for
Understanding Climate Change Impacts

Board on Earth Sciences and Resources

Division on Earth and Life Studies

NATIONAL RESEARCH COUNCIL
OF THE NATIONAL ACADEMIES

THE NATIONAL ACADEMIES PRESS
Washington, D.C.
www.nap.edu

THE NATIONAL ACADEMIES PRESS 500 Fifth Street, N.W. Washington, DC 20001

NOTICE: The project that is the subject of this report was approved by the Governing Board of the National Research Council, whose members are drawn from the councils of the National Academy of Sciences, the National Academy of Engineering, and the Institute of Medicine. The members of the committee responsible for the report were chosen for their special competences and with regard for appropriate balance.

This study was supported by the National Science Foundation under Grant No. EAR-0625247, the U.S. Geological Survey under Award No. 06HQGR0197, and the Chevron Corporation. The opinions, findings, and conclusions or recommendations contained in this document are those of the authors and do not necessarily reflect the views of the National Science Foundation, the U.S. Geological Survey, or the Chevron Corporation.

International Standard Book Number-13: 978-0-309-20915-1
International Standard Book Number-10: 0-309-20915-3
Library of Congress Control Number: 2011930581

Additional copies of this report are available from the National Academies Press, 500 Fifth Street, N.W., Lockbox 285, Washington, DC 20055; (800) 624-6242 or (202) 334-3313 (in the Washington metropolitan area); Internet www.nap.edu

Cover: Cover design by Michael Dudzik.

Printed in the United States of America

THE NATIONAL ACADEMIES

Advisers to the Nation on Science, Engineering, and Medicine

The **National Academy of Sciences** is a private, nonprofit, self-perpetuating society of distinguished scholars engaged in scientific and engineering research, dedicated to the furtherance of science and technology and to their use for the general welfare. Upon the authority of the charter granted to it by the Congress in 1863, the Academy has a mandate that requires it to advise the federal government on scientific and technical matters. Dr. Ralph J. Cicerone is president of the National Academy of Sciences.

The **National Academy of Engineering** was established in 1964, under the charter of the National Academy of Sciences, as a parallel organization of outstanding engineers. It is autonomous in its administration and in the selection of its members, sharing with the National Academy of Sciences the responsibility for advising the federal government. The National Academy of Engineering also sponsors engineering programs aimed at meeting national needs, encourages education and research, and recognizes the superior achievements of engineers. Dr. Charles M. Vest is president of the National Academy of Engineering.

The **Institute of Medicine** was established in 1970 by the National Academy of Sciences to secure the services of eminent members of appropriate professions in the examination of policy matters pertaining to the health of the public. The Institute acts under the responsibility given to the National Academy of Sciences by its congressional charter to be an adviser to the federal government and, upon its own initiative, to identify issues of medical care, research, and education. Dr. Harvey V. Fineberg is president of the Institute of Medicine.

The **National Research Council** was organized by the National Academy of Sciences in 1916 to associate the broad community of science and technology with the Academy's purposes of furthering knowledge and advising the federal government. Functioning in accordance with general policies determined by the Academy, the Council has become the principal operating agency of both the National Academy of Sciences and the National Academy of Engineering in providing services to the government, the public, and the scientific and engineering communities. The Council is administered jointly by both Academies and the Institute of Medicine. Dr. Ralph J. Cicerone and Dr. Charles M. Vest are chair and vice chair, respectively, of the National Research Council.

www.national-academies.org

COMMITTEE ON THE IMPORTANCE OF DEEP-TIME GEOLOGIC RECORDS FOR UNDERSTANDING CLIMATE CHANGE IMPACTS

ISABEL P. MONTAÑEZ, *Chair (2010-2011)*, University of California, Davis
RICHARD D. NORRIS, *Chair (2007-2009)*, Scripps Institution of Oceanography, San Diego, California
THOMAS ALGEO, University of Cincinnati, Ohio
MARK A. CHANDLER, Columbia University, New York
KIRK R. JOHNSON, Denver Museum of Nature and Science, Colorado
MARTIN J. KENNEDY, University of Adelaide, South Australia
DENNIS V. KENT, Rutgers, The State University of New Jersey, Piscataway
JEFFREY T. KIEHL, National Center for Atmospheric Research, Boulder, Colorado
LEE R. KUMP, The Pennsylvania State University, University Park
A. CHRISTINA RAVELO, University of California, Santa Cruz
KARL K. TUREKIAN, Yale University, New Haven, Connecticut

Liaison from the Board on Earth Sciences and Resources

KATHERINE H. FREEMAN, The Pennsylvania State University, University Park

National Research Council Staff

DAVID A. FEARY, Study Director
NICHOLAS D. ROGERS, Research Associate
JENNIFER T. ESTEP, Financial and Administrative Associate
COURTNEY R. GIBBS, Program Associate
ERIC J. EDKIN, Senior Program Assistant

Preface

The drive to better understand how Earth's climate has responded to natural and anthropogenic forcing over the geologically recent past has resulted in a plethora of observational and modeling paleoclimate studies seeking to understand climate dynamics associated with glacial and interglacial cyclicity. From these near-time paleoclimate studies the scientific community has developed a refined understanding of the complex—and often nonlinear—dynamics of the Earth's climate system and has delineated an array of environmental and ecological impacts that have accompanied climate change over the past few thousands to hundreds of thousands of years. The knowledge gleaned from this near-time archive has been of great importance for predicting Earth's immediate future climate. There is, however, a growing appreciation by the scientific community that the changes observed over the past few decades may lead to a degree of warming and associated climate, resource, and ecological changes well beyond those of the "icehouse" climate state in which humans evolved and currently live. Critical insights to understanding such changes are contained in the deep-time geological record that captures the response of the climate system to the full spectrum of internal and external forcings, and their feedbacks, experienced over Earth history.

There is little dispute within the scientific community that humans are changing Earth's climate on a decadal to century timescale. This change, however, is happening within the context of geological time and geological magnitude, and thus must be evaluated and understood at a comparable scale. The fossil fuel CO_2 released to the atmosphere by humans will impact the climate system for tens to hundreds of thousands of years, because of the timescales over which the natural feedbacks in the climate system sweep CO_2 from the atmosphere. Consequently, the changes to the hydrosphere, cryosphere, chemosphere, and biosphere that are occurring now and projected within this century could pale in comparison to those that are possible over the next few centuries. Furthermore, it is studies of deep-time climates that have revealed the nature of the complex network of short- and long-term processes and feedbacks that govern the global climate system. Perhaps more importantly, these studies have also revealed the potential for positive feedbacks that typically operate on century to millennial timescales but may become operative on much shorter, human timescales with continued warming. There are important lessons to be learned from the repeated inability of existing climate models to recreate the large changes in global temperature distributions and climates that can be deciphered from deep-time records of past climates that were much different than Earth's current glacial state. Despite its potential, the deep-time geological record is undertapped scientifically, particularly because it hosts the only records of major, and at times rapid, transitions across climate states as well as records of past warming far greater than witnessed by Earth over the past few tens of millions of years but within the scope of that projected for our future.

This report is the committee's collective assessment of both the demonstrated and the underdeveloped potential of the deep-time geological record to inform us about the dynamics of the global climate system in response to the spectrum of forcings and conditions under which it has operated. A large part of our effort was directed toward describing past climate changes and their impacts on regional climates, water resources, marine and terrestrial ecosystems, and the cycling of life-sustaining elements, emphasizing the lessons that have been learned uniquely from deep-time worlds. While revealing gaps in scientific knowledge of past climate states, we highlight a range of high-priority scientific issues with potential for major advances in the scientific understanding of climate processes. Understanding how the complex network of processes and feedbacks that make up the global climate system operates at various timescales—and over the full range of climate variability experienced through Earth's history—is a high priority for improving projections of future climatic conditions and the impacts on the surface systems that can be anticipated with such change. To that end, we propose a research agenda—and an implementation strategy to address this agenda—that builds on the evolving cross- and interdisciplinary nature of deep-time paleoclimate science.

As the question is increasingly raised as to whether Earth could return—on a human timescale—to a greenhouse climate analogous to that of more than 35 million years ago, it is essential that science thoroughly understands the mechanisms, triggers, and thresholds of climate change in past warmer worlds and the associated magnitudes, rates, and impacts of such change. We hope that the readers of this report share our collective enthusiasm for the richness and societal relevance of the geological record and for the exciting opportunities for enhanced knowledge provided by deep-time paleoclimatology.

The committee would like to thank the National Science Foundation, the U.S. Geological Survey, and Chevron Corporation for enabling this report through their financial support and participation in the study process. As chair, I would also like to acknowledge the committee members for the insights they brought to the table and their efforts in translating a plentitude of knowledge into lucid statements and informative figures. The process was not always linear and involved crossing multiple thresholds; but analogous to climate change, we anticipate that the outcome of the process, this report, will "impact" scientific understanding of the potential of Earth's global climate system for change. Study Director David Feary deserves a special acknowledgment for his tireless effort and patience, aptitude for skillful persuasion, and wonderful wit—it was a privilege to work with him. The committee is also grateful for the support of National Research Council staff members Nicholas Rogers, Courtney Gibbs, and Eric Edkin.

Isabel Patricia Montañez, *Chair*
Committee on the Importance of Deep-Time Geologic Records for Understanding Climate Change Impacts

Acknowledgments

This report was greatly enhanced by those who made presentations to the committee at public committee meetings and by the participants at the open workshop sponsored by the committee to gain community input—David Beerling, Ray Bernor, Karen Bice, Scott Borg, Gabriel Bowen, Robert DeConto, Harry Dowsett, Alexey Federov, Chris Fielding, Margaret Frasier, Linda Gundersen, Bill Hay, Patricia Jellison, H. Richard Lane, Tim Lyons, Thomas Moore, Paul Olsen, Mark Pagani, Judy Parrish, Martin Perlmutter, Chris Poulsen, Greg Ravizza, Dana Royer, Nathan Sheldon, Walt Snyder, Linda Sohl, Lynn Soreghan, Christopher Swezey, Robert Thompson, Thomas Wagner, Debra Willard, Scott Wing, James Zachos, and Richard Zeebe. The presentations and discussions at these meetings provided invaluable input and context for the committee's deliberations. The provision of additional text and figures by Ron Blakey, Paul Olsen, Shanan Peters, Brad Sageman, Lynn Soreghan, Jim Zachos, and Richard Zeebe is also gratefully acknowledged.

This report has been reviewed in draft form by individuals chosen for their diverse perspectives and technical expertise, in accordance with procedures approved by the National Research Council's Report Review Committee. The purpose of this independent review is to provide candid and critical comments that will assist the institution in making its published report as sound as possible and to ensure that the report meets institutional standards for objectivity, evidence, and responsiveness to the study charge. The review comments and draft manuscript remain confidential to protect the integrity of the deliberative process. We wish to thank the following individuals for their participation in the review of this report:

Ken Caldeira, Carnegie Institution of Washington, Stanford, California
Christopher R. Fielding, University of Nebraska, Lincoln
Paul E. Olsen, Lamont-Doherty Earth Observatory, Columbia University, Palisades, New York
Heiko Pälike, National Oceanography Centre, University of Southampton, United Kingdom
Christopher J. Poulsen, University of Michigan, Ann Arbor
Bradley B. Sageman, Northwestern University, Evanston, Illinois
Linda E. Sohl, NASA Goddard Institute for Space Studies, Columbia University, New York
Thomas N. Taylor, University of Kansas, Lawrence
Ellen Thomas, Yale University, New Haven, Connecticut
James C. Zachos, University of California, Santa Cruz

Although the reviewers listed above have provided many constructive comments and suggestions, they were not asked to endorse the conclusions or recommendations nor did they see the final draft of the report before its release. The review of this report was overseen by William W. Hay, University of Colorado Museum, and Peter M. Banks, Executive Office, National Academy of Sciences, Washington, D.C. Appointed by the National Research Council, they were responsible for making certain that an independent examination of this report was carried out in accordance with institutional procedures and that all review comments were carefully considered. Responsibility for the final content of this report rests entirely with the authoring committee and the institution.

Contents

EXECUTIVE SUMMARY 1

SUMMARY 5

1 INTRODUCTION 16
Committee Charge and Scope of This Study, 24

2 LESSONS FROM PAST WARM WORLDS 26
Climate Sensitivity to Increasing CO_2 in a Warmer World, 29
Tropical and Polar Climate Stability and Latitudinal Temperature Gradients in a Warmer World, 33
Hydrological Processes and the Global Water Cycle in a Warmer World, 36
Sea Level and Ice Sheet Fluctuations in a Warmer World, 40
Expansion of Oceanic Hypoxia in a Warmer World, 45
Biotic Response to a Warmer World, 51

3 CLIMATE TRANSITIONS, TIPPING POINTS, AND THE POINT OF NO RETURN 63
Icehouse-Greenhouse Transitions, 65
How Long Will the Greenhouse Last? 76

4 DECIPHERING PAST CLIMATES—RECONCILING MODELS AND OBSERVATIONS 81
Climate Model Capabilities and Limitations, 86
Indicators of Climate Sensitivity Through Time—Proxies for CO_2 and Marine Temperatures, 92
Indicators of Regional Climates, 100
Indicators of Oceanic pH and Redox, 104

5 IMPLEMENTING A DEEP-TIME CLIMATE RESEARCH AGENDA 106
Elements of a High-Priority Deep-Time Climate Research Agenda, 107
Strategies and Tools to Implement the Research Agenda, 113
Education and Outreach—Steps Toward a Broader Community Understanding of Climates in Deep Time, 131

6 CONCLUSIONS AND RECOMMENDATIONS 138
Strategies and Tools to Implement a Deep-Time Climate Research Agenda, 143
Education and Outreach—Steps Toward a Broader Community Understanding of Climates in Deep-Time, 146

REFERENCES 149

APPENDIXES

A Committee Biographical Sketches 181
B Workshop Agenda and Participants 185
C Presentations to Committee 192
D Acronyms and Abbreviations 193

Executive Summary

By the end of this century, without a reduction in emissions, atmospheric CO_2 is projected to increase to levels that Earth has not experienced for more than 30 million years. Critical insights to understanding how Earth's systems would function in this high-CO_2 environment are contained in the records of warm periods and major climate transitions from Earth's geological past.

Throughout its long geological history, Earth has had two fundamentally different climate states—a cool "icehouse" state characterized by the waxing and waning of continental-based ice sheets at high latitudes, and a "greenhouse" state characterized by much warmer temperatures globally and only small—or no—ice sheets. Although Earth has been in an icehouse state throughout the time that humans evolved and for the previous 30 million years, Earth has been in the warmer greenhouse state for most of the past 600 million years of geological time.

As greenhouse gas emissions propel Earth toward a warmer climate state, an improved understanding of climate dynamics in warm environments is needed to inform public policy decisions. Research on the climates of Earth's deep past can address several questions that have direct implications for human civilization: How high will atmospheric CO_2 levels rise, and how long will these high levels persist? Have scientists underestimated the sensitivity of Earth's surface temperatures to dramatically increased CO_2 levels? How quickly do ice sheets decay and vanish, and how will sea level respond? How will global warming affect rainfall and snow patterns, and what will be the regional consequences for flooding and drought? What effect will these changes, possibly involving increasingly acidic oceans and rapidly modified continental climates, have on regional and global ecosystems? Because of the long-lasting effects of this anthropogenic perturbation on the climate system, has permanent change—from a human point of view—become inevitable? How many thousands of years will it take for natural processes to reverse the projected changes?

The importance of these questions to science and to society prompted the National Science Foundation, the U.S. Geological Survey, and Chevron Corporation to commission the National Research Council to describe the existing understanding of Earth's past climates, and to identify focused research initiatives to better understand the insights that the deep-time record offers into the response of Earth systems to projected future climate change. Throughout this report, "deep time" refers to that part of Earth's history that must be reconstructed from rock, and is older than historical or ice core records. Although the past 2 million years of the Pleistocene are included in "deep time," most of the focus of the research described or called for in this report is on the long record of Earth's history prior to the Pleistocene.

Although deep-time greenhouse climates are not exact analogues for the climate of the future, past warm climates—and particularly abrupt global warming events—provide important insights into how physical, biogeochemical, and biological processes operate under warm conditions. These insights particularly include the role of greenhouse gases in causing—or "forcing"—global warming; the impact of warming on ice sheet stability, sea level, and on oceanic and hydrological processes; and the consequences of global warming for ecosystems and the global biosphere. As Earth continues to warm, it may be approaching a critical climate threshold beyond which rapid and potentially permanent—at least on a human timescale—changes may occur, prompting major societal questions: How soon could abrupt and dramatic climate change occur, and how long could such change persist?

HIGH-PRIORITY DEEP-TIME CLIMATE RESEARCH AGENDA

The following six elements of a deep-time scientific research agenda have the potential to address enduring scientific issues and produce exciting and critically important results over the next decade:

- To understand how sensitive climates are to increased atmospheric CO_2.
- To understand how heat is transported around the globe and the controls on pole-to-equator thermal gradients.

- To understand sea level and ice sheet stability in a warm world.
- To understand how water cycles will operate in a warm world.
- To understand abrupt transitions across tipping points into a warmer world.
- To understand ecosystem thresholds and resilience in a warming world.

STRATEGIES AND TOOLS TO IMPLEMENT THE RESEARCH AGENDA

Implementing the deep-time paleoclimate research agenda described above will require four key infrastructure and analytical elements:

- Development of additional and improved estimates of precipitation, seasonality, aridity, and soil productivity in the geological past.
- Continental and ocean drilling transects to collect high-resolution records of past climate events and transitions, to determine climate parameters before and after these events, and to model the dynamic processes causing these transitions.
- Paleoclimate modeling focusing on past warm worlds and extreme and/or abrupt climate events, at high resolution to capture regional paleoclimate variability. Model outputs will be compared with climate records from drilling transects and fine-tuned.
- A transition from single-researcher or small-group research efforts to a much broader-based interdisciplinary collaboration of observation-based scientists with climate modelers for team-based studies of important paleoclimate events.

ENCOURAGING A BROADER COMMUNITY UNDERSTANDING OF CLIMATES IN DEEP TIME

The public—and indeed many scientists—have minimal appreciation of the value of understanding deep-time climate history and appear largely unaware of the relevance of far distant past times for Earth's future. The paleoclimate record contains surprising facts—there have been times when the poles were forested rather than being icebound; there were times when the polar seas were warm; there were times when tropical forests grew at midlatitudes; more of Earth history has been greenhouse than icehouse. Such straightforward concepts provide an opportunity to help disparate audiences understand that the Earth has archived its climate history and that this archive, while not fully understood, is perhaps science's best tool to understand Earth's climate future.

The possibility that our world is moving toward a "greenhouse" future continues to increase as anthropogenic carbon builds up in the atmosphere, providing a powerful motivation for understanding the dynamics of Earth's past "greenhouse" climates that are recorded in the deep-time geological record. An integrated research program—a deep-time climate research agenda—to provide a considerably improved understanding of the processes and characteristics over the full range of Earth's potential climate states offers great promise for informing individuals, communities, and public policy.

Summary

By the end of this century, without a reduction in emissions, atmospheric CO_2 is projected to increase to levels that Earth has not experienced for more than 30 million years. Critical insights to understanding how Earth's systems would function in this high-CO_2 environment are contained in the records of warm periods and major climate transitions from Earth's geological past.

Earth is currently in a cool "icehouse" state, a climate state characterized by continental-based ice sheets at high latitudes. When considering the immense expanse of geological time, an icehouse Earth has not been the norm; for most of its geological history Earth has been in a warmer "greenhouse" state. As increasing levels of atmospheric CO_2 drive Earth toward a warmer climate state, an improved understanding of how climate dynamics could change is needed to inform public policy decisions. Research on the climates of Earth's deep past can address several questions that have direct implications for human civilization: How high will atmospheric CO_2 levels rise, and how long will these high levels persist? Have scientists underestimated the sensitivity of Earth surface temperatures to dramatically increased CO_2 levels? How quickly do ice sheets decay and vanish, and how will sea level respond? How will global warming affect rainfall and snow patterns, and what will be the regional consequences for flooding and drought? What effect will these changes, possibly involving increasingly acidic oceans and rapidly modified continental climates, have on regional and global ecosystems? Because of the long-lasting effects of this anthropogenic perturbation on the climate system, has permanent change—from a human point of view—become inevitable? How many thousands of years will it take for natural processes to reverse the projected changes?

The importance of these questions to science and to society prompted the National Science Foundation, the U.S. Geological Survey, and Chevron Corporation to commission the National Research Council (NRC) to report on the present state of scientific understanding of Earth's geological record of past climates, and to identify focused research initiatives to better understand the insights offered by Earth's deep-time record into the response of Earth systems to potential future climate change. Throughout this report, the "deep time" geological record refers to that part of Earth's history that is older than historical or ice core records, and therefore must be reconstructed from rocks. Although the past 2 million years of the Pleistocene are included in "deep time," most of the research described or called for in this report focuses on the long record of Earth's history prior to the Pleistocene.

The study of climate history and processes during the current glacial (icehouse) state shows the sensitivity of Earth's climate system to various external and internal factors and the response of key components of the Earth system to such change. The resolution and high precision of these datasets capture environmental change on human timescales, thereby providing a critical baseline against which future climate change can be assessed. However, this more recent paleoclimate record captures only part of the known range of climate phenomena; the waxing and waning of ice sheets in a bipolar glaciated world under atmospheric CO_2 levels at least 25 percent lower than current levels.

In contrast to this reasonably well documented record of recent climate dynamics and at least partial understanding of the short-term feedbacks that have operated in icehouse states of the recent past, the climate dynamics of past periods of global warming and major climate transitions are considerably less well understood. Paleoclimate records offer potential for a much improved understanding of the long-term equilibrium sensitivity of climate to increasing atmospheric CO_2 and of the impact of global warming on atmospheric and ocean circulation, ice sheet stability and sea level response, ocean acidification, regional hydroclimates, and the health of marine and terrestrial ecosystems. Deep-time paleoclimate records uniquely offer the temporal continuity required to understand how both short-term (decades to centuries) and long-term (millennia to tens of millennia) climate system feedbacks have played out over the longer periods of time in Earth's history. This understanding will provide valuable insights on how Earth's climate would respond to, and recover from, the levels of greenhouse gas forcing that are projected to occur if no efforts are made to reduce emissions.

Although deep-time greenhouse climates are not exact analogues for the climate of the future, past warm climates—and in particular abrupt global warming events—could provide important insights into how physical, biogeochemical, and biological processes operate under warm conditions. These insights particularly include the role of greenhouse gases in causing—or "forcing"—global warming; the impact of warming on ice sheet stability, on sea level, and on oceanic and hydrological processes; and the consequences of global warming for ecosystems and the global biosphere.

As Earth continues to warm, it may be approaching a critical climate threshold beyond which rapid and potentially permanent—at least on a human timescale—changes not anticipated by climate models tuned to modern conditions may occur. Components of the climate system that are particularly vulnerable to being forced across such thresholds by increasing atmospheric CO_2 include the following: the loss of Arctic summer sea ice, the stability of the Greenland and West Antarctic Ice Sheets, the vigor of Atlantic thermohaline circulation, the extent of Amazon and boreal forests, and the variability of El Niño-Southern Oscillation. The deep-time geological archive of climate change concerning such thresholds could provide insight to several major societal questions—How soon could abrupt and dramatic climate change occur, and how long could such change persist?

The deep-time geological record provides several examples of such climate transitions, perhaps the most dramatic of which—with potential parallels to the near future—was the Paleocene-Eocene Thermal Maximum. This abrupt greenhouse gas-induced global warming began ~55 million years ago with the large and rapid releases of "fossil" carbon and major disruption of the carbon cycle. Global warming was accompanied by extreme changes in hydroclimate and accelerated weathering, deep-ocean acidification, and possible widespread oceanic anoxia. Whereas regional climates in the mid- to high latitudes became wetter and were characterized by increased extreme precipitation events, other regions, such as the western interior of North America, became more arid. With this intense climate change came ecological disruption, with the appearance of modern mammals (including primates), large-scale floral and faunal ecosystem migration, and widespread extinctions in the deep ocean.

A key requirement for an improved understanding of deep-time climate dynamics is the integration of high-resolution observational records across critical intervals of paleoclimate transition with more sophisticated modeling. Projections of climate for the next century are based on general climate models (GCMs) that have been developed and tuned using records of the "recent" past. In part, this reflects the high levels of radiometric calibration and temporal resolution (subannual to submillennial) offered by near-time paleoclimate archives. A critical prerequisite for accurate projections of future regional and global climate changes based on GCMs, however, is that these models use parameters that are relevant to the potential future climate states we seek to better understand. In this context, the recent climate archive captures only a small part of the known range of climate phenomena because it has been derived from a time dominated by low and relatively stable atmospheric CO_2 and bipolar glaciations. Modeling efforts will have to be expanded to capture the full range of variability and climate-forcing feedbacks of the global climate system, in particular for the past "extreme climate events" and warmer Earth intervals that may serve as analogues for future climate. An additional requirement is the need to improve the more comprehensive Earth system models to better capture regional climate variability, particularly with different boundary conditions (e.g., paleogeography, paleotopography, atmospheric pCO_2 [partial pressure of carbon dioxide], solar luminosity).

HIGH-PRIORITY DEEP-TIME CLIMATE RESEARCH AGENDA

The committee, with substantial contributions from a broad cross section of the scientific community, has identified the following six elements of a deep-time scientific research agenda as having the highest priority to address enduring scientific issues and produce exciting and critically important results over the next decade:

Improved Understanding of Climate Sensitivity and CO_2-Climate Coupling

Determining the sensitivity of the Earth's mean surface temperature to increased greenhouse gas levels in the atmosphere is a key requirement for estimating the likely magnitude and effects of future climate change. The paleoclimate record, which captures the climate response to a full range of radiative forcing, can uniquely contribute to a better understanding of how climate feedbacks and the amplification of climate change have varied with changes in atmospheric CO_2 and other greenhouse gases. A critical scientific focus is the determination of long-term equilibrium climate sensitivity on multiple timescales, in particular during periods of greenhouse gas forcing comparable to that anticipated within and beyond this century. Using the deep-time geological archive to address these questions will require focused efforts to improve the accuracy and precision of existing proxies, together with efforts to develop new proxies for past atmospheric pCO_2, surface air and ocean temperatures, non-greenhouse gases, and atmospheric aerosol contents. With these improved data, a hierarchy of models can be used to test how well processes other than

CO_2 forcing (e.g., non-CO_2 greenhouse gases; solar and aerosol forcing) can explain anomalously warm and cold periods and to critically evaluate the degree to which feedbacks and climate responses are characteristic of greenhouse gas forcing.

Climate Dynamics of Hot Tropics and Warm Poles

Although paleoclimate reconstructions indicate that the tropics have been much warmer in the past and that anomalous polar warmth characterized some warm paleoworlds, the current understanding of how tropical and higher-latitude temperatures respond to increased CO_2 forcing remains limited. This reflects the mismatch between modern observational data, which have been used to define the hypothesis of thermostatic regulation of tropical surface temperatures, and climate model simulations run with large radiative forcings that argue against tropical climate stability. The current consensus is that tropical ocean temperatures seem not to have been regulated by such a tropical thermostat. Notably, the deep-time record indicates that the mechanisms and feedbacks in the modern icehouse climate system, which have controlled tropical temperatures and a high pole-to-equator thermal gradient, may not straightforwardly apply in warmer worlds, suggesting that additional feedbacks probably operated under warmer mean temperatures. An improved understanding of these processes, which may drive further changes in surface temperatures in a future warmer world, is important in light of the substantial effects that higher temperatures would have on tropical ecosystems and ultimately on regional extratropical climates through teleconnections.

Sea Level and Ice Sheet Stability in a Warm World

Large uncertainties in the theoretical understanding of ice sheet dynamics currently hamper the ability to predict future ice sheet and sea level response to continued climate forcing. For example, paleoclimate studies of intervals within the current icehouse climate state document variations in the extent of ice sheet coverage that cannot be reproduced by state-of-the-art coupled climate-ice sheet models. These studies further indicate that equilibrium sea level in response to current warming may be substantially higher than model projections. Efforts to address these issues will have to focus on past periods of ice sheet collapse that accompanied transitions from icehouse to greenhouse conditions, in order to provide context and understanding of the "worst-case" forecasts for the future. Such studies will also refine scientific understanding of long-term equilibrium sea level change in a warmer world, the nature of ice age termination, and the timescales at which such feedbacks and responses could operate in the future. An integral component of such studies should be a focus on improving the ability to separate the temperature and seawater signals recorded in biogenic marine proxies, including refinement of existing paleotemperature proxies and the development of new geochemical and biomarker proxies.

Understanding the Hydrology of a Hot World

Studies of past climates and climate models strongly suggest that the greatest impact of continued CO_2 forcing would be regional climate changes, with consequent modifications of the quantity and quality of water resources—particularly in drought-prone regions—and impacts on ecosystem dynamics. A fundamental component of research to understand hydrology under warmer conditions is the requirement—because of its potential for large feedbacks to the climate system—for an improved understanding of the global hydrological cycle over a full spectrum of CO_2 levels and climate conditions. The deep-time record uniquely archives the processes and feedbacks that influence the hydrological cycle in a warmer world, including the effect of high-latitude unipolar glaciation or ice-free conditions on regional precipitation patterns in lower latitudes. Understanding these mechanisms and feedback processes requires the collection of linked marine-terrestrial records that are spatially resolved and of high temporal resolution, precision, and accuracy.

Understanding Tipping Points and Abrupt Transitions to a Warmer World

Studies of past climates show that Earth's climate system does not respond linearly to gradual CO_2 forcing, but rather responds by abrupt change as it is driven across climatic thresholds. Modern climate is changing rapidly, and there is a possibility that Earth will soon pass thresholds that will lead to even larger and/or more rapid changes in its environments. Climate system behavior whereby a small change in forcing leads to a large change in the system represents a "tipping phenomenon" and the threshold at which an abrupt change occurs is the "tipping point." The question of how close Earth is to a tipping point, when it could transition into a new climate state, is of critical importance. Because of their proven potential for capturing the dynamics of past abrupt changes, intervals of abrupt climate transitions in the geological record—including past hyperthermals—should be the focus of future collaborative paleoclimate, paleoecologic, and modeling studies. Such studies should lead to an improved understanding of how various components of the climate system responded to abrupt transitions in the past, in particular during times when the rates of change were sufficiently fast to imperil diversity. This research also will help determine whether there exist thresholds and feedbacks in the climate system of which we are unaware—especially in warm worlds and past icehouse-to-greenhouse transitions. Moreover, targeting such intervals for more detailed investigation is a critical requirement for constraining how long any abrupt climate change might persist.

Understanding Ecosystem Thresholds and Resilience in a Warming World

Both ecosystems and human society are highly sensitive to abrupt shifts in climate because such shifts may exceed the tolerance of organisms and, consequently, have major effects on biotic diversity, human investments, and societal stability. Modeling future biodiversity losses and biosphere-climate feedbacks, however, is inherently difficult because of the complex, nonlinear interactions with competing effects that result in an uncertain net response to climatic forcing. How rapidly biological and physical systems can adjust to abrupt climate change is a fundamental question accompanying present-day global warming. An important tool to address this question is to describe and understand the outcome of equivalent "natural experiments" in the deep-time geological record, particularly where the magnitude and/or rates of change in the global climate system were sufficiently large to threaten the viability and diversity of many species, which at times led to mass extinctions. The paleontological record of the past few million years does not provide such an archive because it does not record catastrophic-scale climate and ecological events. As with the other elements of a deep-time research agenda, improved dynamic models, more spatially and temporally resolved datasets with high precision and chronological constraint, and data-model comparisons are all critical components of future research efforts to better understand ecosystem processes and dynamic interactions.

STRATEGIES AND TOOLS TO IMPLEMENT THE RESEARCH AGENDA

Implementing the deep-time paleoclimate research agenda described above will require four key infrastructure and analytical elements:

- Development of additional and improved quantitative estimates of paleoprecipitation, paleoseasonality, paleoaridity, and paleosoil conditions (including paleoproductivity). This will require targeted efforts to refine existing proxies and develop new proxies[1] in particular where the level of precision and accuracy—and thus the degree of uncertainty in inferred climate parameter estimates—can be significantly reduced. Proxy improvement efforts should include strategies for better constraining the paleogeographic setting of proxy records, including latitude and altitude or bathymetry, as well as the development of proxies for greenhouse gases other than CO_2 (e.g., methane). Ultimately, such new and/or refined mineral and organic proxies will permit the collection of multiproxy paleoclimate time series that are spatially resolved, temporally well constrained, and of high precision and accuracy.

- A transect-based deep-time drilling program designed to identify, prioritize, drill, and sample key paleoclimate targets—involving a substantially expanded continental drilling program and additional support for the existing scientific ocean drilling program—to deliver high-resolution, multiproxy archives for the key paleoclimate targets across terrestrial-paralic-marine transects and latitudinal or longitudinal transects. Such a drilling strategy will permit direct comparison of the marine and terrestrial proxy records that record fundamentally different climate responses—local and regional effects on continents compared with homogenized oceanic signals.
- Enhanced paleoclimate modeling with a focus on past warm worlds and extreme and/or abrupt climate events, including improved parameterization of conditions that are relevant to future climate, development of higher-resolution modules to capture regional paleoclimate variability, and an emphasis on paleoclimate model intercomparison studies and "next-generation" paleoclimate data-model comparisons. An increase in model spatial resolution will be required to capture smaller-scale features that are more comparable to the highly spatially resolved geological data that can be obtained through ocean and continental drilling. Downscaling techniques using either statistical approaches or nested fine-scale regional models, will be required to better compare simulated climate variables to site-specific observational data. Dedicated computational resources for model development and the application of models to specific time periods are important requirements to address this element of the research agenda.
- An interdisciplinary collaboration infrastructure to foster broad-based collaborations of observation-based scientists and modelers. This collaboration will allow team-based studies of important paleoclimate time slices, incorporating climate and geochemical models; will expand

[1] When used in the scientific sense (rather than the more common context of stock ownership or voting delegation), proxies refer to parameters that "stand in" for other parameters that cannot be directly measured. For example, tree ring width is commonly considered to be primarily a proxy for past temperature.

capabilities for the development, calibration, and testing of highly precise and accurate paleoclimate proxies; and will allow the continued development of digital databases to store proxy data and facilitate multiproxy and record comparisons across all spatial and temporal scales. Such broad-based and interdisciplinary cultural and technological infrastructure will require acceptance and endorsement by both the scientific community and the funding agencies that support deep-time paleoclimatology and paleobiology-paleoecology studies. Making the transition from single-researcher or small-group research efforts to a much broader-based interdisciplinary collaboration will be possible only through a modification of scientific research culture, and will require substantially increased programmatic and financial support.

EDUCATION AND OUTREACH STRATEGIES— STEPS TOWARD A BROADER COMMUNITY UNDERSTANDING OF CLIMATES IN DEEP TIME

Despite the potential of the deep-time geological record to provide unique insights into the global climate system's sensitivity, response, and ability to recover from perturbation, the public—and indeed many scientists—have minimal appreciation of the value of understanding deep-time climate history and appear largely unaware of the relevance of far distant times for Earth's future. The deep-time climate research community has not made the point strongly enough that the record of the past can be inspected both for insights into the Earth's climate system and for making more informed projections for the future. A strategy for education and outreach to convey the lessons contained within deep-time records should be tailored to several specific audiences:

- K-12 elementary and secondary education, with teachers and children requiring different strategies. Museums are a key resource for educating children, providing access to the inherently interesting dinosaur story as a window into deep time. The involvement of teachers in scientific endeavors (e.g., the teacher-at-sea program of the Integrated Ocean Drilling Program) provides opportunities to demystify science and convey the excitement of scientific discovery, as well as disseminating scientific information.
- Colleges and universities, where distinguished lecture tours and summer schools can add to the more traditional learning elements in geoscience courses. The integration of deep-time paleoclimatology into environmental science curricula offers an additional opportunity to convey the relevance of the deep-time record.
- To involve and educate the general public, the deep-time obser-

vation and modeling communities have opportunities to break into the popular science realm by emphasizing their more compelling and understandable elements. Immediate opportunities exist for the popularization of ice cap and ocean drilling research, both of which occur in dramatic settings that are unfamiliar and interesting to the general public. These illustrate "science in action," showing scientists undertaking interesting activities in the pursuit of knowledge.

- Potential scientific collaborators from the broader climate science community can obtain an increased understanding of the potential offered by paleoclimate data and modeling through the creation or use of forums in which scientists from different disciplines exchange information and perspectives. This is effectively done *within* disciplines by talks and symposia at national disciplinary meetings and *between* disciplines at meetings of broader groups, such as those hosted by the American Association for the Advancement of Science.
- Policy makers require scientifically credible and actionable data on which to base their policies. Faced with a diversity of opinions, they need credible sources of information. This report and other NRC reports attempt to play this role, but in a much broader sense the scientific community must strive to make the presentation of deep-time paleoclimate information as understandable as possible.

The paleoclimate record contains facts that are surprising to most people: for example, there have been times when the poles were forested rather than being icebound; there were times when the polar seas were warm, and there were times when tropical forests grew at midlatitudes. For the majority of Earth's history, the planet has been in a greenhouse state rather than in the current icehouse state. Such concepts provide an opportunity to help disparate audiences understand that the Earth has archived its climate history and that this archive, while not fully understood, provides crucial lessons for improving our understanding of Earth's climate future. Such relatively simple but relevant messages provide a straightforward mechanism for an improved understanding in the broader community of the importance of paleoclimate studies.

The possibility that our world is moving toward a "greenhouse" future continues to increase as anthropogenic carbon builds up in the atmosphere, providing a powerful motivation for understanding the dynamics of Earth's past "greenhouse" climates that are recorded in the deep-time geological record. It is the deep-time climate record that has revealed feedbacks in the climate system that are unique to warmer worlds—and thus are not archived in more recent paleoclimate records—and might be expected to

become increasingly relevant with continued warming. It is the deep-time record that has revealed the thresholds and tipping points in the climate system that have led to past abrupt climate change, including amplified warming, substantial changes in continental hydroclimate, catastrophic ice sheet collapse, and greatly accelerated sea level rise. Also, it is uniquely the deep-time record that has archived the full temporal range of climate change impacts on marine and terrestrial ecosystems, including ecological tipping points. An integrated research program—a deep-time climate research agenda—to provide a considerably improved understanding of the processes and characteristics over the full range of Earth's potential climate states offers great promise for informing individuals, communities, and public policy.

1

Introduction

The atmosphere's concentration of carbon dioxide—a potent greenhouse gas—has been increasing in recent years faster than had been forecast by even the most extreme projections of a decade ago. At current carbon emission rates, Earth will experience atmospheric CO_2 levels within this century that have not occurred since the warm "greenhouse" climates of more than 34 million years ago. Atmospheric pCO_2 could reach as high as 2000 ppmv[1] if fossil fuel emissions remain unabated, all fossil fuel resources are used, and carbon sequestration efforts remain at present-day levels (Kump, 2002; Caldeira and Wicket, 2003). As oceanographer Roger Revelle noted more than 50 years ago, humans are launching an uncontrolled "Great Geophysical Experiment" with the planet to observe how burning fossil fuels will affect all aspects of the climate, chemistry, and ecology of Earth (Revelle and Suess, 1957). Despite the high stakes for humans and the natural environment that will result from forcing an "icehouse" planet into "greenhouse" conditions, we still have only a poor idea of what this rapidly approaching greenhouse world will be like. However, studies of past climate states do provide a vision of this climate future and the substantial and abrupt (years to decades) climate shifts that are likely to usher in these changed climate conditions.

This projected rise in atmospheric CO_2 levels—perhaps at unprecedented rates—raises a series of major questions with direct implications for human civilization:

[1] The atmospheric concentration of carbon dioxide—the partial pressure of the gas (pCO_2)—is expressed in units of ppmv (parts per million by volume).

- What is the sensitivity of air and ocean (both shallow and deep) temperatures to dramatically increased CO_2 levels?
- How high will atmospheric CO_2 levels rise, and for how long will these high levels persist?
- How quickly do ice sheets decay and vanish, and consequently how rapidly does sea level change? Also, if the Arctic is to become permanently ice-free, how will this affect thermohaline circulation and regional and global climate patterns?
- Are there processes in the climate system that are not currently apparent or understood that will become important in a warmer world?
- How will global warming affect rainfall and snow levels, and what will be the regional consequences for flooding and drought?
- What effect will these changes have on the diversity of marine biota? What will be the impact on—and response of—terrestrial ecosystems?
- Has climate change become inevitable? How long will it take to reverse the projected changes through natural processes?

How Earth's climate system has responded to past episodes of increasing and elevated atmospheric CO_2 is a critical element of the answers to these questions.

Temperature Response to Increasing CO_2

Recent syntheses suggest that climate sensitivity—the response of global mean surface temperature to a doubling of atmospheric CO_2 levels—lies between 1.5 and 6.2°C (Hegerl et al., 2006; IPCC, 2007; Hansen et al., 2008). The lower end of this range (≤3°C) is based on modern data and paleoclimate records extending back no further than the Last Glacial Maximum of 20,000 years ago, and therefore these estimates factor in only the short-term climate feedbacks—such as water vapor, sea ice, and aerosols—that operate on subcentennial timescales. Climate sensitivity, however, is likely to be enhanced under higher atmospheric CO_2 and significantly warmer conditions due to long-term positive feedbacks that typically are active on much longer timescales (thousands to tens of thousands of years) (Hansen et al., 2008; Zachos et al., 2008; Pagani et al., 2010). These physical and biochemical feedbacks—such as changes in ice sheets and terrestrial biomes as well as greenhouse gas release from soils and from methane hydrates in tundra and ocean sediments—however, may become increasingly more relevant on human timescales (decades) with continued global warming (Hansen and Sato, 2001; Hansen et al., 2008). Determining the deep-time record of *equilibrium* climate sensitivity—in particular during periods of elevated CO_2 and at timescales at which long-term climate feedbacks operate—is thus a critical element in evaluating climate theories more thoroughly and for constraining the magnitude and effects of future temperature increases (Box 1.1).

Alternating Icehouse and Greenhouse—Earth's Climate History

Earth is currently in an "icehouse" state—a climate state characterized by continental-based ice sheets at high latitudes. Human evolution took place in this bipolar (i.e., with ice sheets at each pole) icehouse (NRC, 2010), and civilizations arose within its most recent interglacial phase. Such icehouse states, however, account for far less of Earth's history than "hothouse" states (Figure 1.2).

Most paleoclimate studies have focused on the interglacial-glacial cycles that have prevailed during the past 2 million years of the current icehouse, to link instrumental records with geological records of the recent past and to exploit direct records of atmospheric gases preserved in continental glaciers. These relatively recent (Pleistocene) records document systematic fluctuations in atmospheric greenhouse gases in near concert with changes in continental ice volume, sea level, and ocean temperatures. Their decadal- to millennial-scale resolution has improved scientific understanding of the complex climate dynamics of the current bipolar glacial state, including the ability of climate to change extremely rapidly—in some cases over a decade or less (Taylor et al., 1993; Alley et al., 2003). Perhaps most importantly, recent ice core archives reveal that during the past 800,000 years—prior to the industrial rise in pCO_2—the current icehouse has been characterized by atmospheric CO_2 levels of less than 300 parts per million (Siegenthaler et al., 2005).

In contrast to this reasonably well documented record of recent climate dynamics and at least partial understanding of the short-term (sub-centennial) feedbacks that have operated in icehouse states of the near past, scientific understanding of the climate dynamics for past periods of global warming—when Earth was in a "greenhouse" climate state—is much less advanced. The paleoclimate records of deep-time worlds,[2] however, are the closest analogue to Earth's anticipated future climate—one that will be warmer and greenhouse gas forced beyond that experienced in the past 2 million years, as atmospheric CO_2 contents have already surpassed by about 35 percent those that applied during the Pleistocene glacial-interglacial cycles. This deep-time geological archive records the full spectrum of Earth's climate states and uniquely captures the ecosys-

[2] The deep-time geological record that is the subject of this report refers to that part of Earth's history that must be reconstructed from rocks, older than historical or ice core records. Although the past 2 million years of the Pleistocene are included in deep time, most of the focus of the research described or advocated here is the long record of Earth's history prior to the Pleistocene.

BOX 1.1
Societal Effects—
What Do the Projected Temperature Changes Really Mean?

Global temperatures are projected to rise by at least 1°C, and perhaps up to 6°C (Figure 1.1), by the end of this century (IPCC, 2007). The human consequences of this steep rise in greenhouse gases are likely to be substantial, with decreased precipitation in already drought-prone regions and widespread social, economic, and health effects (IPCC, 2007). One yardstick to better appreciate these effects is to consider the roughly 0.2-0.5°C rise in global temperatures that accompanied the Medieval Warm Period at ~1000 A.D. This modest rise in temperatures resulted in meadows and stunted beech forests in fjords in southwest Greenland, as well as ice-free shipping lanes that allowed Vikings to colonize Greenland between 982 A.D and 1400 A.D. Drought throughout the Americas and Southeast Asia, coincident with this warming event, has been invoked as a contributing factor in the collapse of the Anasazi, the Classic Mayan, the South American Moche civilization, and the Khmer empire of Angkor Wat (e.g., Haug et al., 2003; Hodell et al., 2005; Ekdahl et al., 2008; Zhang et al., 2008).

Fluctuations in average global temperatures during the glacial-interglacial cycles of the past several hundred thousand years caused major shifts in the areal extent of continental ice sheets and greater than 100-m sea level changes, with some interglacial periods up to 2-3°C warmer than the present day (Otto-Bliesner et al., 2006). Large-scale changes in carbon cycling and overall greenhouse gas contents, including 50 percent variations in atmospheric CO_2, occurred *in response* to interglacial warmings (Sigman and Boyle, 2000; Lea, 2004; Siegenthaler et al., 2005), highlighting the potential for amplification of future CO_2-driven global warming through climate-CO_2 feedbacks. In fact, estimates of temperature response to all modern forcings, including human and naturally induced factors, indicate the potential for ~0.6-1.4°C of additional warming—with no additional greenhouse gas forcing—as the long-term feedbacks that typically operate on thousands to tens of thousands of years (e.g., changes in surface albedo feedback with variation in ice sheet and vegetation coverage) become operative on human timescales (Hansen et al., 2008). The substantial societal impacts from past temperature increases that were of lesser magnitude than those anticipated during this century raise obvious questions about the societal impacts that are likely to result from future temperature rise.

continued

BOX 1.1 Continued

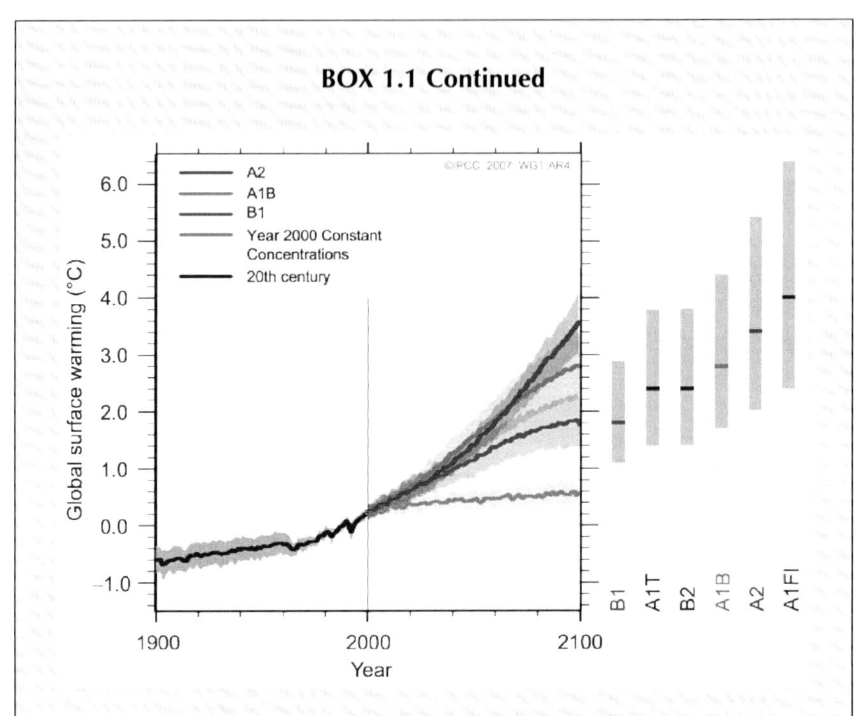

FIGURE 1.1 Projections (colored lines), with uncertainty bounds of ±1 standard deviation (shading), for future surface temperature rise from models that use different economic scenarios. Scenario A2 represents "business as usual" where temperature is projected to rise by the end of the century between 2° an
d 5.5°C if no effort is made to constrain the rise of CO_2 levels. The solid bars at right indicate the best estimate (solid line) and possible ranges (shading) for each scenario.
SOURCE: IPCC (2007, Figure SPM.5, p. 14).

tem response to, and interaction with, this full range of climate changes. The deep-time record thus offers the potential for a much improved understanding of the long-term equilibrium sensitivity of climate to increasing pCO_2, and of the impact of major climate change on atmospheric and ocean circulation; ice sheet stability and sea level response; ocean acidification and hypoxia; regional hydroclimates; and the diversity, radiation, and decline of marine and terrestrial organisms (see Box 1.2). Furthermore, the deep-time paleoclimate records uniquely offer the temporal continuity required to understand how both short- (subcentennial) and long-term (millennial-scale) climate system feedbacks have played out over the

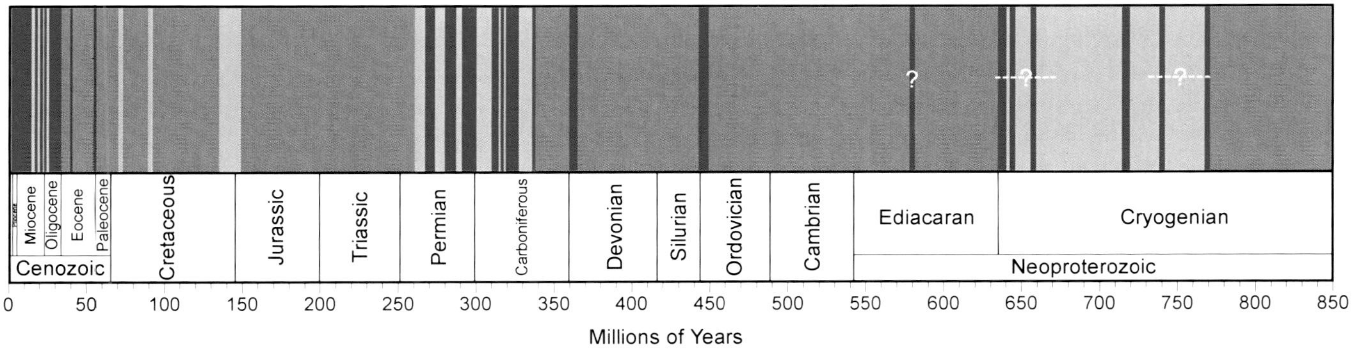

FIGURE 1.2 Although warmer greenhouse conditions (red-brown intervals) have dominated most of the past ~1 billion years of Earth's history, there have been extended periods of cool "icehouse" conditions (light-blue intervals) including intervals for which there is evidence of continental ice sheets at one or both poles (shown as darker blue bars). The question marks in the Cryogenian reflect uncertainties associated with the geographic extent and duration of inferred glacial events during this time (Allen and Etienne, 2008; Kendall et al., 2009). The Paleocene-Eocene Thermal Maximum and Mid-Eocene Thermal Maximum are shown as red bars. The current icehouse began ~34 million years ago with increased glaciation in Antarctica and accelerated with northern hemisphere glaciation over the past 3 million years.
SOURCES: Compiled based on Miller et al. (2003); Montañez et al. (2007), Bornemann et al. (2008); Brezinski et al. (2008); Fielding et al. (2008); Zachos et al. (2008); and Macdonald et al. (2010).

BOX 1.2
The Nonlinear Development of Life on Earth

The Earth has been populated with life, as we know it today, through a protracted and nonlinear history of evolution characterized by repeated extinctions and radiations (Figure 1.3). The biosphere in the Precambrian was dominated by single-cellular organisms such as microbes and cyanobacteria, capable of building massive reefal structures and living in the overall oxygen-poor conditions of the oceans of the time. In contrast, Phanerozoic faunal life—the past 542 million years—has been characterized by a metazoan fauna, rich in diversity, that arose following the geologically rapid radiation of life in the latest Precambrian (see Figure 1.2 for timescale). Floral ecosystems of equal diversity soon populated much of the Earth following their evolution ~450 million years ago. Throughout Earth history, physiological evolution and ecosystem dynamics have been intricately linked to various surface processes and systems (e.g., landscapes, ocean and atmospheric composition and circulation, soil and hydrological processes) through interactions and feedbacks, examples of which are presented in this report and are a fundamental component of interdisciplinary deep-time studies. Earth's deep-time history offers numerous examples of how ecosystems, geosystems, and climate systems have operated in the absence of various major groups of life and under conditions far more extreme than those of the present day—time intervals when oceans lacked the major elements of their current buffering capacity or were hypoxic, when the poles lacked ice sheets, and/or when atmospheric CO_2 levels were higher by hundreds to thousands of parts per million of volume.

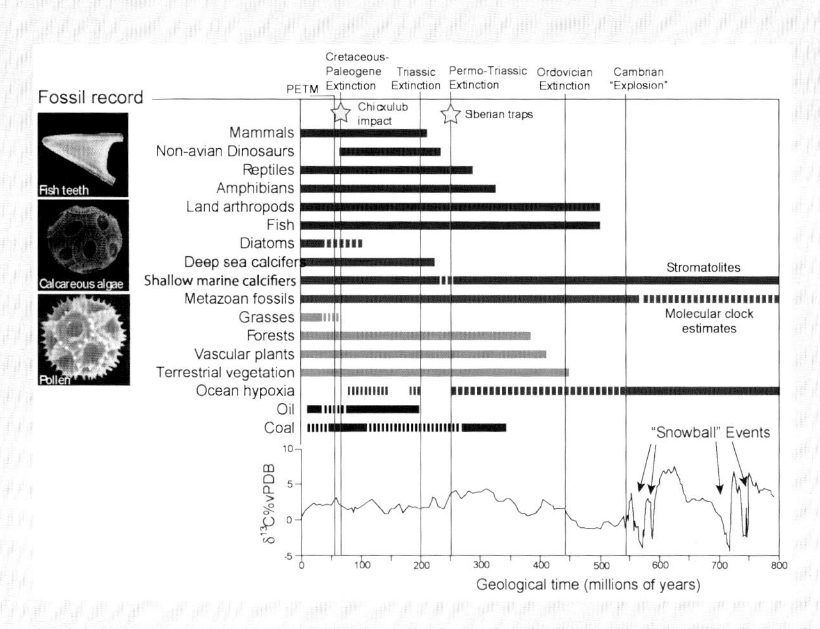

FIGURE 1.3 Timing of major events in late Precambrian and Phanerozoic evolution. From bottom to top: record of the carbon cycle from carbon isotopes, showing the transition from the high-amplitude cycles of the late Neoproterozoic—including several "snowball Earth" episodes—to the much more muted trends of the Phanerozoic (Hayes et al., 1999). Periods of abundant coal and oil formation, which include the extensive coal units of the Carboniferous and Permian, Cretaceous coal, and extensive coal deposits of the Paleogene Arctic, as well as oil deposits formed during Jurassic and Cretaceous oceanic anoxic events and Mio-Pliocene oil deposits of the Pacific Rim (Windley, 1995), are shown in black. Ocean hypoxia (red line) illustrates the reduction in the extent of anoxic or hypoxic conditions in the deep sea with time, with low oxygen common in the Paleozoic and intermittent episodes of basinwide to global oceanic anoxic events in the Mesozoic. The extent of vegetation cover is shown with green lines, and major groups of oceanic organisms that contribute to global geochemical cycles either through burrowing (metazoans; Sheehan, 2001), marine calcifiers that buffer ocean pH (Ridgwell et al., 2003), or diatoms with their role in the silica cycle (Ridgwell et al., 2002; Cortese et al., 2004; Lazarus et al., 2009), are shown in blue. Fish (brown line) appear in the early Cambrian (Shu, 1999) and give rise to terrestrial amphibians in the Devonian (Selden, 2001). The invasion of land is accomplished by terrestrial arthropods well before the appearance of terrestrial vertebrates. Non-avian dinosaurs and mammals evolved from reptiles in the early Mesozoic (Sereno, 1999; Brusatte et al., 2008). Reptiles show multiple invasions of the oceans in the Mesozoic, and mammalian groups invade the ocean several times in the Cenozoic.

longer periods of time (millennia to hundreds of thousands of years) that are necessary to fully understand how Earth's climate responds to, and recovers from, the levels of greenhouse gas forcing that will result from fossil fuel burning over the next century.

COMMITTEE CHARGE AND SCOPE OF THIS STUDY

The National Science Foundation, U.S. Geological Survey, and Chevron Corporation, with input from the Geosystems initiative[3] and the broader research community, commissioned the National Research Council to describe the present state of understanding of Earth's geological record of past climates, as well as to identify focused research initiatives that would enhance the understanding of this record and thereby improve predictive capabilities for the likely parameters and impacts of future climate change. The study committee was also charged to present advice on research implementation and public outreach strategies (Box 1.3).

To address this charge, the National Research Council assembled a committee of 12 members with broad disciplinary expertise; committee biographical information is presented in Appendix A. The committee held four meetings between February 2008 and February 2009, convening in Washington, D.C.; Boulder, Colorado; and twice in Irvine, California. The major focal point for community input to the committee was a 2-day open workshop held in May 2008 (see Appendix B), where concurrent breakout sessions interspersed with plenary addresses enabled the committee to gain a thorough understanding of community perspectives regarding the status of existing research as well as future research priorities. Additional briefings by sponsors and keynote addresses from other speakers were presented at the initial meeting of the committee (see Appendix C).

The paleoclimate archive contained in the geological record both offers an opportunity and assigns a responsibility for Earth and climate science to effectively predict what is likely to happen as Earth warms and to offer projections with enough precision to assist society to mitigate and/or adapt to future changes. The examination of climate states in the deep-time geological record has the potential to provide unique information about how Earth's climate dynamics operate over long time frames and during changes of large magnitude. Earth's pending transition into warmer climates provides the motivation for the description of the understanding of past warm periods presented in Chapter 2, and the transitions into and out of different climate states over differing timescales is the

[3] The Geosystems initiative is an interdisciplinary, community-based initiative focused on understanding the wealth of "alternative-Earth" climatic extremes archived in older parts of the geological record, as the basis for understanding Earth's climate future. See http://www.geosystems.org/.

BOX 1.3
Statement of Task

The geologic record contains physical, chemical, and biological indicators of a range of past climate states. As recent changes in atmospheric composition cause Earth's climate to change, and amid suggestions that future change may cause the Earth to transition to a climatic state that is dramatically different from that of the recent past, there is an increasing focus on the geologic record as a repository of critical information for understanding the likely parameters and impacts of future change. To further our understanding of past climates, their signatures, and key environmental forcing parameters and their impact on ecosystems, an NRC study will:

- Assess the present state of knowledge of Earth's deep-time paleoclimate record, with particular emphasis on the transition periods of major paleoclimate change.
- Describe opportunities for high-priority research, with particular emphasis on collaborative multidisciplinary activities.
- Outline the research and data infrastructure that will be required to accomplish the priority research objectives.

The report should also include concepts and suggestions for an effective education and outreach program.

focus of Chapter 3. The capabilities and limitations of existing models and proxies used to describe and understand past climates are addressed in Chapter 4, providing the backdrop for the recommendations for a high-priority deep-time climate research agenda and strategies to implement this agenda which are contained in Chapter 5. Some elements of this report—particularly the descriptions of existing scientific understanding of the paleoclimate record and the processes that have controlled that record—are necessarily technical; nevertheless, every effort has been made to present the material in terms that are accessible to the broadest possible audience.

2

Lessons from Past Warm Worlds

Alfred Wegener's concept of continental drift, reformulated in the modern theory of plate tectonics, arose in part as a way to explain the geographic distribution of paleoclimate indicators in ancient rocks. Permo-Carboniferous (~300-million-year-old) glacial deposits in distinctly nonpolar regions of present-day Africa, South America, and Australia rectify to polar latitudes when the ancient supercontinent of Gondwana is reconstructed. Continental drift transported them through broad climate belts—humid tropics, arid subtropics, moist and cool temperate zones, and cold and arid polar regions (Box 2.1). Paleoclimate reconstructions, however, reveal that although paleogeography and the plate tectonics that control continental configurations are important, they are not the major determinant of climate change. Global warm climates have prevailed when large continents covered the poles, and deep "snowball Earth" glaciations occurred when there apparently were no polar continents. Instead, it appears that the greenhouse gas content of the atmosphere was the key factor in determining whether a particular interval of Earth's past was an icehouse or a greenhouse.

Although most deep-time greenhouse climates occurred when there were distinctly different continental configurations, and thus are not direct analogues for the future, past warm climates and abrupt transitions into even hotter states (known as hyperthermal events; Thomas et al., 2000) provide important insights into how physical, biogeochemical, and biological processes operate under warm conditions more analogous to what is anticipated for the future than the moderate and stable climates of the Holocene (past 10,000 years) or the relatively warm interglacials of

BOX 2.1
Continental Drift and Climate

Plate tectonics has been rearranging Earth's configuration of continents ever since the plates on Earth became rigid approximately 2.5 billion years ago (Figure 2.1). On long (millions of years) timescales, the movement of tectonic plates—and the continents that ride upon them—has strongly influenced Earth's distribution of solar insolation, ocean and atmospheric circulation, and carbon cycling between the Earth's deep and shallow reservoirs, thereby profoundly impacting global climate, sea level, and the overall planetary ecology.

The arrangement of the continents through time is most reliable for the past 800 million years, the period for which the chronostratigraphic tools necessary for reconstructions are available. The global views presented in Figure 2.1 show how the continents on Earth's surface may have appeared during four intervals of time that are noted throughout this report: the unipolar glaciated Pennsylvanian (300 million years ago [Ma]), mid-Cretaceous (105 Ma), Eocene (50 Ma), and mid-Pliocene (3 Ma).

The major transitions between climatic icehouse and greenhouse conditions are ultimately most probably driven by the deep Earth processes of plate tectonics, as a function of the long-term balance between CO_2 degassing at spreading centers and the conversion of atmospheric CO_2 to mineral carbon through long-term silicate weathering and oceanic carbonate formation (Berner, 2004). For example, the eruptions of large igneous provinces in the mid-Cretaceous and the subduction of the carbonate-rich tropical Tethys Sea in the early Cenozoic are the most likely cause of the high-CO_2 equilibrium climates of the Cretaceous and Eocene greenhouses. Conversely, uplift of the Himalayas and Tibetan Plateau associated with "docking" of the Indian subcontinent with Asia (~40 Ma), and the evolution of vascular land plants in the early Paleozoic (~450 Ma), led to the sequestration of atmospheric CO_2 through enhanced weathering of silicate minerals (Ruddiman, 2007; Archer, 2009).

continued

BOX 2.1 Continued

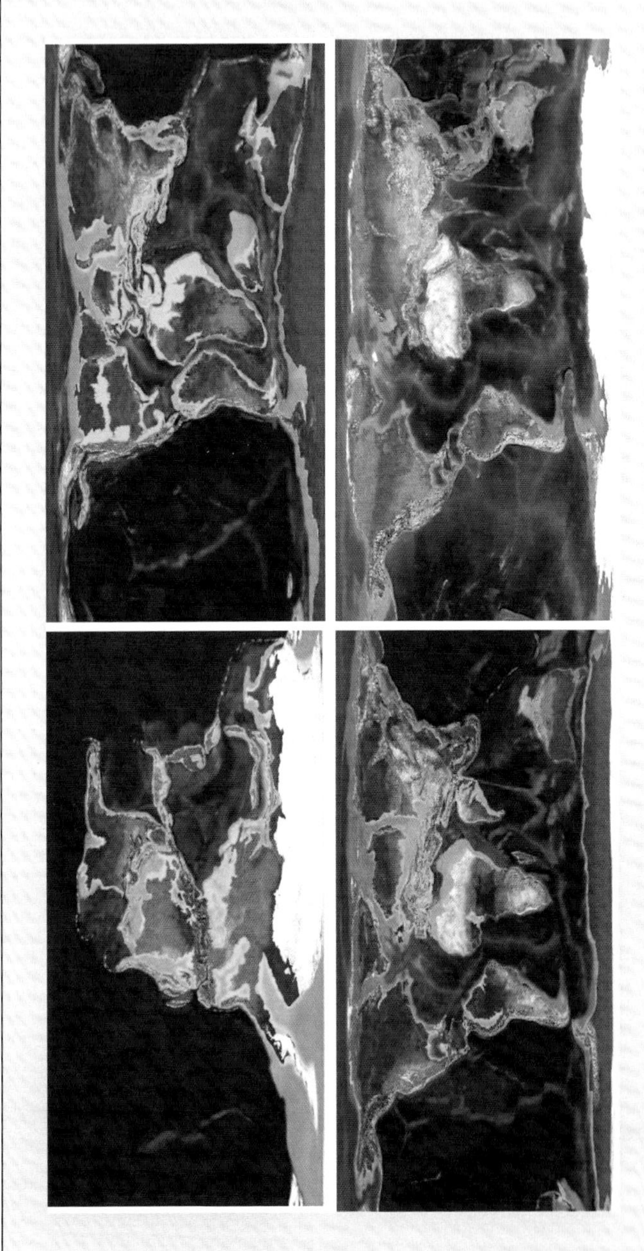

FIGURE 2.1 Continental configurations for the Pennsylvanian (upper left), mid-Cretaceous (upper right), Eocene (lower left), and mid-Pliocene (lower right). Topography was defined on the basis of digital elevation maps of modern Earth from the U.S. Geological Survey; colors portray climate and vegetation distribution based on a synthesis of all geological literature relevant to each time slice.
SOURCE: Courtesy R.C. Blakey, Colorado Plateau Geosystems.

the Pleistocene (past 2 million years). The following sections describe the insights provided by understanding past warm periods, including the role of greenhouse gases in controlling—or "forcing"—global warming; the impact of warming on ice sheet stability, sea level, and oceanic and hydrological processes; and the consequences of global warming for ecosystems and the global biosphere.

CLIMATE SENSITIVITY TO INCREASING CO_2 IN A WARMER WORLD

Fundamentally, Earth's climate results from the balance between absorbed energy from the sun and radiant energy emitted from Earth's surface, with changes to either component resulting in a forcing of the climate system. The net forcing of the climate system over geological time caused episodes of warming and cooling that are coincident with greenhouse and icehouse climates, respectively. Most projections indicate that, by the end of this century, climate forcing resulting from increased CO_2 will be at least of the same magnitude as that experienced in the early Cenozoic (during the late Eocene, ~34 Ma) (Figure 2.1), and possibly analogous to estimates for the Cretaceous Period (~80-120 Ma)—probably one of the times of greatest radiative forcing since the evolution of animals (Hay, 2010).

Climate sensitivity—the equilibrium warming resulting from a doubling of atmospheric carbon dioxide relative to preindustrial levels of CO_2—provides a measure of how the climate system responds to external forcing factors and is also used to compare global climate model outputs to understand why different models respond to the same external forcings with different outputs. Climate sensitivity to CO_2 strongly influences the magnitude of warming that Earth will experience at any particular time in the future (Box 2.2). The magnitude of climate sensitivity and Earth's surface temperature are determined by a myriad of short-term (human timescales) and long-term (thousands to tens of thousands) interactions and feedbacks (e.g., water vapor, cloud properties, sea ice albedo, snow albedo, ice sheet and terrestrial biome distribution, ocean-atmosphere CO_2 interaction, and silicate weathering).

As noted above, synthesis of the various estimates of Earth's climate sensitivity for the past 20,000 years has lead to the general conclusion that sensitivity most probably lies in the range of 1.5 to 4.5°C (IPCC, 2007), with some recent projections suggesting that the value may be even as high as 6-8°C (Hansen et al., 2008; Knutti and Hegerl, 2008). However, estimates of equilibrium climate sensitivity averaged over tens to hundreds of millennia (i.e., long term) and extending back for 400 million years are minimally between 3 and 6°C (Royer et al., 2007). For the most recent period of global

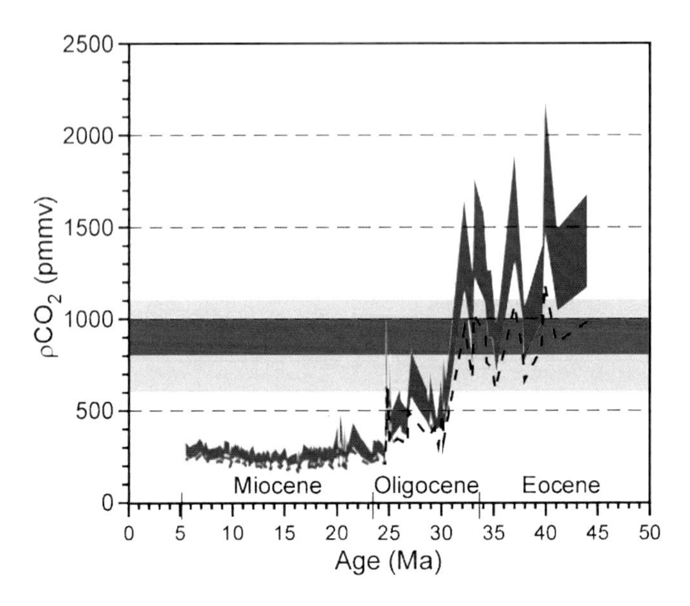

FIGURE 2.2 Estimated atmospheric pCO_2 for the past 45 million years (late Eocene through Miocene) calculated using all available stable carbon isotopic values of diunsaturated alkenones in deep-sea sediments. Values of CO_{2aq} were translated to atmospheric pCO_2 using Henry's Law and a range of dissolved phosphate values and sea surface temperatures for each site, and a salinity of 35 parts per thousand. The dark gray shaded region shows the range of maximum to intermediate estimates, and the dashed line represents minimum estimates. The uncertainty in pCO_2 estimates ranges from ~20 percent for the Miocene to 30 to 40 percent for the Paleogene. The broad pale red band (pCO_2 values of 600-1,100 parts per million by volume) encompasses most of the CO_2 concentration range for nonmitigation emission scenarios projected for the end of this century (figure 10.26 in IPCC, 2007); the dark red band (values of 800-1,000 ppmv) corresponds to the Intergovernmental Panel on Climate Change (IPCC) A2 "business-as-usual" scenario.
SOURCE: Modified after Pagani et al. (2005).

warming, the middle Pliocene (~3.0-3.3 Ma), climate sensitivity may have been as high as 7-9.6°C ± 1.4°C per CO_2 doubling (Pagani et al., 2010). Such values, which are well above short-term climate sensitivity estimates based on more recent paleoclimate and instrumental records, indicate that long-term feedbacks operating at accelerated timescales (decadal to centennial) promoted by global warming can substantially magnify an initial temperature increase.

As Earth moves toward a warmer climate state, it is important to understand the extent to which climate sensitivity will change due to processes

BOX 2.2
Why Does Climate Sensitivity Matter?

For any particular increase in atmospheric CO_2 (and other greenhouse gases), a system with high climate sensitivity to CO_2 will warm more in the future than a world with low climate sensitivity. Thus, if the climate sensitivity is high, restricting future global warming will require a larger reduction in future CO_2 emissions than if climate sensitivity is lower. Comparison of emission scenarios for the period until 2100, calculated for a range of CO_2 stabilization targets (Figure 2.3A) and the corresponding *equilibrium* global average temperature increases (Figure 2.3B; IPCC, 2007), based on the Intergovernmental Panel on Climate Change (IPCC) range of climate sensitivities (2 to 4.5°C), illustrates the impact of fossil carbon emissions on future surface temperatures and the extent of reductions required to limit the warming to ≤2°C relative to preindustrial conditions.

Even if anthropogenic carbon emissions to the atmosphere are reduced, CO_2 levels will continue to increase for a century or more because the removal of CO_2 from the atmosphere by natural processes of carbon sequestration (e.g., CO_2 absorption by the surface ocean, CO_2 fertilization of terrestrial vegetation) is slow (Archer et al., 2009). Consequently, temperature increases may continue for several centuries until equilibrium temperatures are reached, especially for higher CO_2 stabilization targets. However, even if climate sensitivity is at the lower end of the possible range, global temperature increases of ≥2°C will be reached with CO_2 stabilization levels of 450-550 parts per million by volume. Given that *equilibrium* temperature increases may be protracted, emissions could continue to increase into the middle of this century (Figure 2.3; Caldeira et al., 2003). However, if the climate sensitivity is 4.5°C or greater, then a significant and immediate reduction in CO_2 emissions—to levels ultimately below those of the present day—is required to stay below a target warming of 2°C.

continued

BOX 2.2 Continued

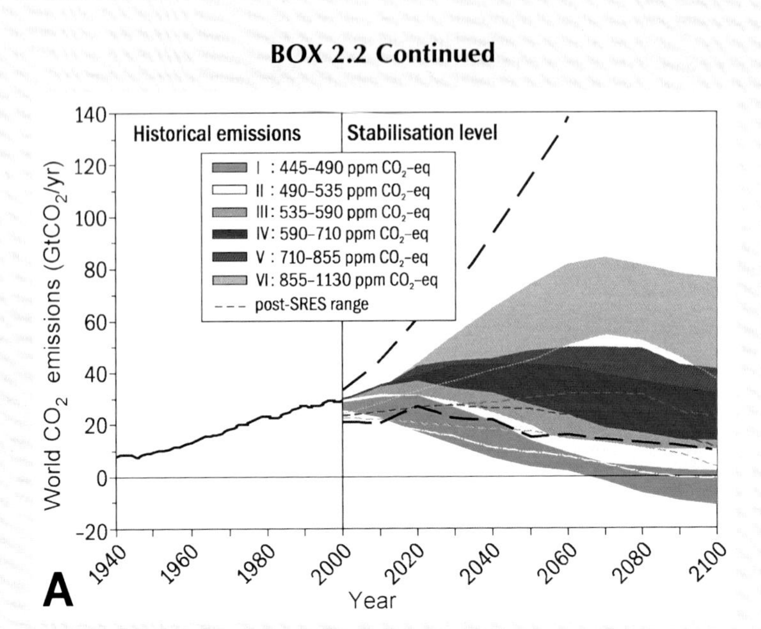

FIGURE 2.3 Global CO_2 emissions and equilibrium global average temperature increases for a range of target CO_2 stabilization levels. (A) Measured (1940 to 2000) and projected (colored shading; 10th to 90th percentile) global CO_2 emissions for the range of IPCC emission scenarios and associated stabilization CO_2 levels indicated by roman numerals (ppm CO_2-eq). (B) Corresponding relationship between the different

that have not operated in recent icehouse climate regimes. One simple example is to consider a warm world with no sea ice at either pole—as CO_2 increases, the sea ice albedo feedback is removed and therefore this negative feedback's contribution to climate sensitivity is absent. In addition, it is important to determine the potential for nonlinear responses that are specific to a greenhouse or transitional world, and whether such responses would enhance climate sensitivity. For example, the destabilization of continental ice sheets resulting from warming of polar regions can potentially lead to a decrease in deep-water formation, thereby affecting global ocean circulation, stratification, and carbon cycling, leading to higher climate sensitivity than indicated by present estimates. In sufficiently warm climates, even water vapor has a nonlinear dependence on temperature, and this can introduce new and potentially rapid feedbacks, operating at a subdecadal scale, into the climate system. Destabilization of methane and its release

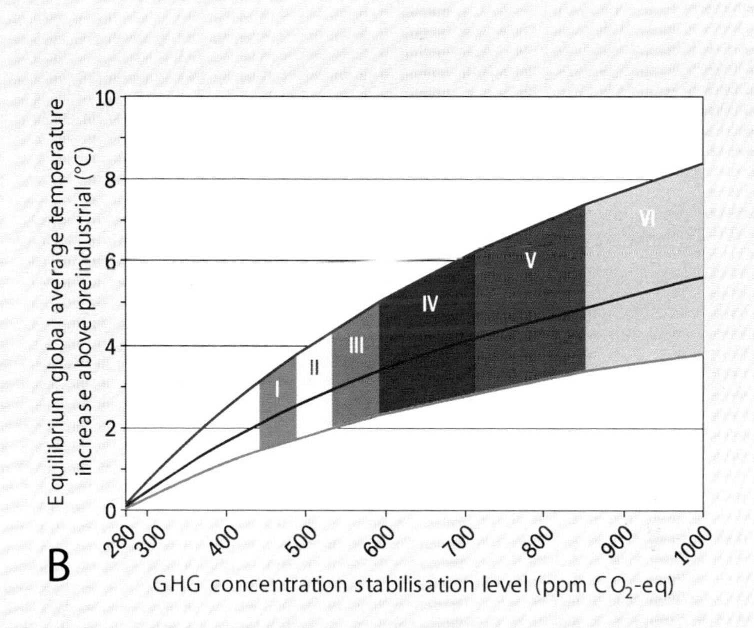

CO_2 stabilization targets shown in (A) and equilibrium global average temperature increase above preindustrial levels. Colored regions for each stabilization target were calculated for a range of climate sensitivity (2–4.5°C) and "best estimate" climate sensitivity of 3°C (blue solid line in middle of shaded area).
SOURCE: IPCC (2007, Figure 5.1, page 66).

into the atmosphere in response to warming— through either the melting of terrestrial permafrost reservoirs or the dissolution of subseafloor clathrate deposits—would dramatically increase greenhouse gas contents in the atmosphere. Hence, an initial warming from greenhouse gases released by burning fossil fuels could end up releasing even more greenhouse gases from natural sources, exacerbating the original warming of the atmosphere.

TROPICAL AND POLAR CLIMATE STABILITY AND LATITUDINAL TEMPERATURE GRADIENTS IN A WARMER WORLD

With more than half of Earth's surface lying within 30° latitude of the equator, the response of tropical climates to increased greenhouse gas forcing is critically important. Modern observational data (Ramanathan and

Collins, 1991) suggest that western tropical Pacific sea surface temperatures rarely exceed ~30-32°C, and this has led to speculation that the Earth's tropics have a "thermostat" that limits maximum sea surface temperatures. Explanations about how such a thermostat might work have included the buildup of clouds that reflect heat back into space (Ramanathan and Collins, 1991), evaporative cooling (Hartmann and Michelsen, 1993; Pierrehumbert, 1995), winds, or an increase in transport of heat out of the tropics by ocean currents (Clement et al., 1996; Sun and Liu, 1996). Most of these studies have used present-day data to explain surface temperature regulation, although there are artifacts in these datasets that call into question the robustness of the observed trends (Clement et al., 2010). Paleoclimate reconstructions of tropical temperatures during past greenhouse times, however, document sea surface temperatures that were much warmer than modern tropical maxima—possibly as high as 42°C—and thus were probably not thermostatically "regulated" (Bice et al., 2006; Came et al., 2007; Pearson et al., 2007; Trotter et al., 2008; Kozdon et al., 2009).

The discovery of a giant Paleocene snake fossil in South America (Head et al., 2009; Huber, 2009; although see discussions by Makarieva et al., 2009; Sniderman, 2009), as well as other terrestrial paleotemperature indicators such as paleoflora leaf-margin analysis and stable isotope compositions of biogenic apatites and soil minerals (Fricke and Wing, 2004; Tabor and Montañez, 2005; Passey et al., 2010), further suggests anomalously high continental temperatures (~30-34°C) for the terrestrial tropics of past warmer worlds. Additionally, coupled climate model simulations with large radiative forcings and/or paleoclimate simulations for elevated greenhouse gases do not produce a thermostatic regulation of tropical temperatures (e.g., Boer et al., 2005; Poulsen et al., 2007b; Cherchi et al., 2008), suggesting that the tropical warming in response to greenhouse gas forcing is neither moderated nor local in its impacts (Xie et al., 2010). Such deep-time paleoclimate studies have documented that tropical surface temperatures during past greenhouse periods were not thermostatically regulated by the negative feedback processes that operate in the current icehouse climate system, further illustrating how knowledge of deep-time warm periods is fundamental to understanding Earth's climate system.

There is also abundant evidence for anomalous polar warmth during past greenhouse periods (e.g., middle Cretaceous to Eocene, Pliocene; see Figure 2.4) associated with reduced equator-to-pole temperature gradients (e.g., Huber et al., 1995; Crowley and Zachos, 2000; Hay, 2010; Miller et al., 2010). To date, climate models have not been able to simulate this warmth without invoking greenhouse gas concentrations that are notably higher than proxy estimates (Figure 2.4; Bice et al., 2006). This has prompted modeling efforts to explain high-latitude warmth through vegetation (DeConto et al., 1999), clouds (Sloan and Pollard, 1998; Abbot and Tziperman, 2008;

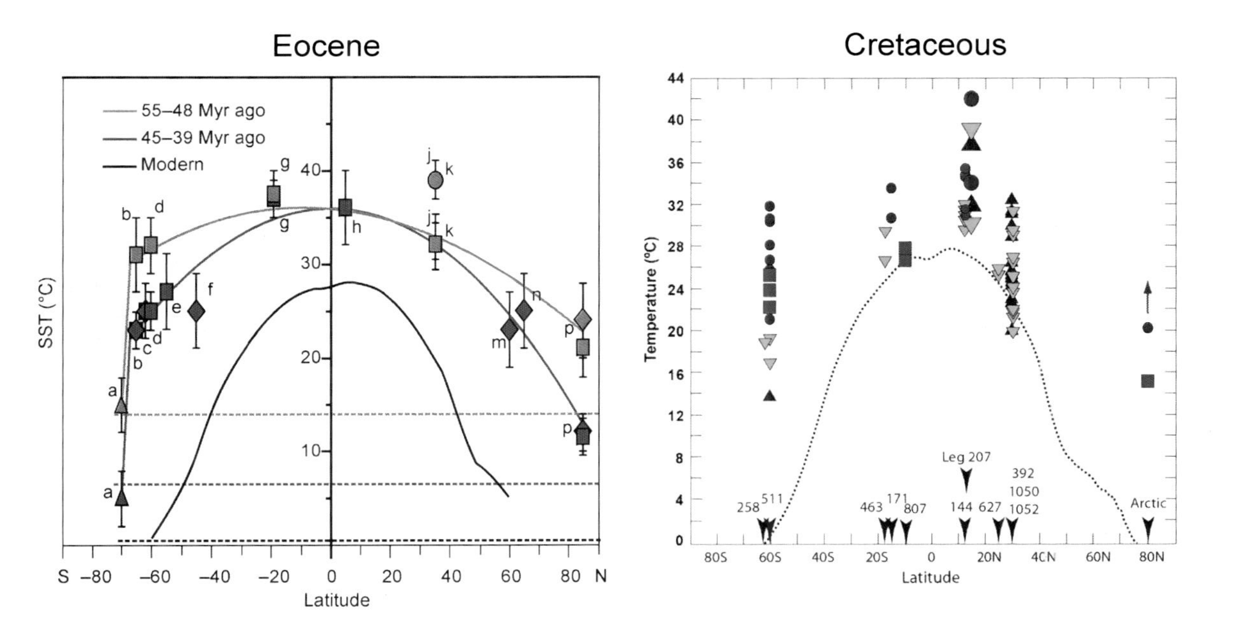

FIGURE 2.4 Zonal mean surface temperature (°C) as a function of latitude for the Eocene and Cretaceous. For the Eocene, the solid lines represent second-order polynomials excluding triangle data, with error bars representing the range of variation. Dashed lines represent deep-sea temperatures. For the Cretaceous, the dotted line is the modern observed zonal surface temperature, while the symbols indicate a compilation of empirical data for specific periods.
SOURCES: Eocene—modified from Bijl et al. (2009), reprinted by permission of Macmillan Publishers Ltd.; Cretaceous—courtesy K. Bice, personal communication, 2010.

Kump and Pollard, 2008), intensified heat transport by the oceans (Barron et al., 1995; Korty et al., 2008), and increased tropical cyclone activity (Sriver and Huber, 2007; Fedorov et al., 2010). The ability to successfully model a reduced latitudinal temperature gradient state, including anomalous polar warmth, presents a first-order check on the efficacy of climate models as the basis for predicting future greenhouse conditions.

Since significant changes in tropical and polar surface temperatures and pole-to-equator temperature gradients occurred in the past, and could occur in a future warmer world, it is imperative to understand the mechanisms and feedbacks that lead to such changes and their consequences for atmospheric and oceanic circulation (Hay, 2008). The fundamental mismatch between model outputs, modern observations, and paleoclimate proxy records discussed above, however, may indicate some very important deficiencies in scientific knowledge of climate and the construction of climate models (e.g., Huber, 2008). Resolution of this disparity, as well as an improved understanding of the anthropogenic signal in observational data, can likely be obtained by analysis of paleoclimate records from past warm worlds.

HYDROLOGICAL PROCESSES AND THE GLOBAL WATER CYCLE IN A WARMER WORLD

Earth's hydrological processes—including precipitation, evaporation, and surface runoff—are susceptible to, and play a critical role in, both past and future climate change (Pierrehumbert, 2002). Large-scale atmospheric processes determine the general position of climate zones and the intensity of precipitation and storms; the intertropical convergence zone is a region of significant rainfall, while large regions of atmospheric subsidence lead to dry desert regions. Regional hydroclimates, such as the Southwest Indian and the East Asian summer monsoons, which affect nearly half of Earth's human population, are highly sensitive to distal climate changes and to mean warming (Sinha et al., 2005; Wang et al., 2005) via teleconnections (e.g., changes in high-latitude surface temperatures or Arctic sea ice extent impact lower-latitude climate through atmospheric processes). Overall, the vapor-holding capacity of the atmosphere increases substantially with increased global mean temperatures if there is no change in the relative humidity. Consequently, climate models for global warming predict an intensified hydrological cycle and, on a global scale, enhanced precipitation (IPCC, 2007).

Observations over the past few decades indicate that precipitation has increased faster (~7 percent per degree of surface warming; Wentz et al., 2007) than that predicted by models (1-3 percent per degree of surface warming; Zhang et al., 2007). Although the reasons for this substantial discrepancy are not understood, one possibility is that the observational data trends are too short to detect long-period changes in evaporation (Wentz et al., 2007). Moreover, analyses of future global warming model simulations (IPCC, 2007) predict that the atmosphere will hold more water as the latitudinal extent of the Hadley cell expands (Held and Soden, 2006; Lu et al., 2007).[1] Warming will also enhance precipitation in the tropics and midlatitudes and will expand subtropical desert regions.

Small changes in tropical sea surface temperatures can impart large changes in global climate patterns, affecting wind strength, relative rates of precipitation versus evaporation, and surface temperatures (Cane, 1998). With evidence from deep-time records for warmer tropical oceans during past periods of global warming, the sensitivity of the climate system to such change is of critical importance for projecting how regional and global climate patterns may change in the future. For example, the Pacific ocean-atmosphere system and its coupled instability, El Niño-Southern Oscillation (ENSO), are maintained by dynamic feedbacks that are sensitive to external forcings (Huber and Caballero, 2003; Cane, 2005). Perturbation of this regional system with continued warming could modulate global climate change, including a shift toward a permanent rather than intermittent El Niño-like state. Climate simulations of future transient global warming, however, offer inconsistent and uncertain projections regarding ENSO behavior (Cane, 2005; Collins, 2005), although predictions of the Intergovernmental Panel on Climate Change (IPCC, 2007) include a change in tropical climate to an El Niño-dominated state. Existing observational data of global warming over the last century are too short term to resolve the relative importance of ocean versus atmospheric feedbacks (Vecchi et al., 2008). In contrast, proxy time series of sufficient continuity, together with complementary climate models of past sustained warm periods illustrate that small, long-term changes in Pacific sea surface temperatures can have a substantial effect on ENSO phenomena, as well as planetary albedo, regional rainfall, and increased atmospheric levels of water vapor—a powerful greenhouse gas (see Box 2.3). If the IPCC (2007) predictions of a long-term increase in El Niño frequency and intensity with continued warming are correct, then the climate impacts could be far-reaching and include amplification of global mean temperatures, in particular in the extratropical regions, as well as widespread drought in some regions coincident with catastrophic flooding in others. The geological records from warm periods, such as the Pliocene, will be key elements for testing these predictions.

Deep-time paleoclimate studies provide a critical perspective, docu-

[1] The Hadley cell is the circulation cell dominating the tropical atmosphere, with rising motion near the equator, poleward flow 10-15 km above the surface, descending motion in the subtropics, and equatorward flow near the surface.

BOX 2.3
Persistent El Niño-like Conditions of the Early Pliocene Warming

The early Pliocene (4.5 to 3 Ma) was characterized by atmospheric pCO_2 levels that were elevated in comparison to preindustrial times, but similar to those of the present day (Pagani et al., 2010; Seki et al., 2010). Globally averaged temperatures were ~3°C higher (Figure 2.5), the northern polar region was ice-free, and a mean state resembling El Niño-like conditions persisted in the Pacific (Wara et al., 2005; Fedorov et al., 2006; Ravelo et al., 2006). This permanent El Niño-like state contributed significantly to overall warming and major changes in the hydrological cycle, including effects in distal regions such as eastern equatorial Africa where perturbations in rainfall patterns may have influenced the evolution of hominins and other vertebrates (deMenocal, 1995; NRC, 2010). Recent deep-time modeling studies have shown that such El Niño-like mean conditions in the Pliocene had far-reaching effects, influencing climate well outside of the tropics (e.g., Shukla et al., 2009).

The persistent El Niño conditions led to weaker Walker circulation, the east-west directed zonal atmospheric circulation. As temperatures in the eastern equatorial region cooled gradually after ~2 Ma, the equatorial sea surface temperature gradient increased dramatically (post-1.8 Ma on Figure 2.6), ultimately establishing strong Walker circulation and the present-day ENSO

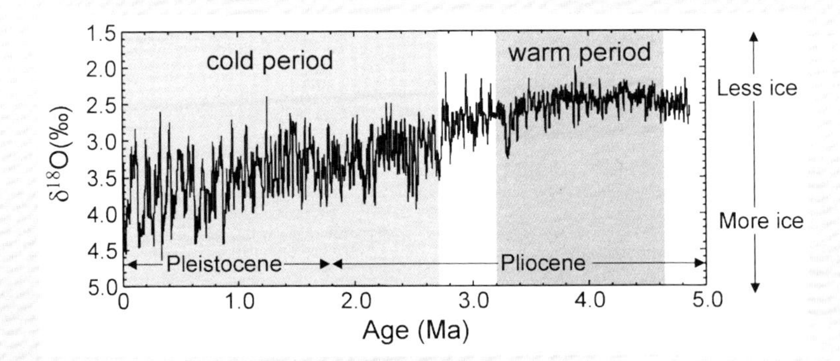

FIGURE 2.5 Benthic foraminifer $\delta^{18}O$ record of high-latitude climate (primarily ice volume) from Ocean Drilling Program (ODP) cores, showing the relative warmth during the early Pliocene, followed by the transition into the cooler later Pliocene and Pleistocene.
SOURCE: Ravelo and Wara (2004).

state. Ocean-atmosphere coupled model simulations for the Pliocene have not successfully reproduced the observed persistent El Niño mean state, with close to modern temperatures in the west and warmer than modern in the east. Furthermore, climate simulations and paleoclimate data for the Eocene greenhouse period indicate that despite much higher global temperatures, El Niño variability was comparable to that of the present day (Huber and Caballero, 2003).

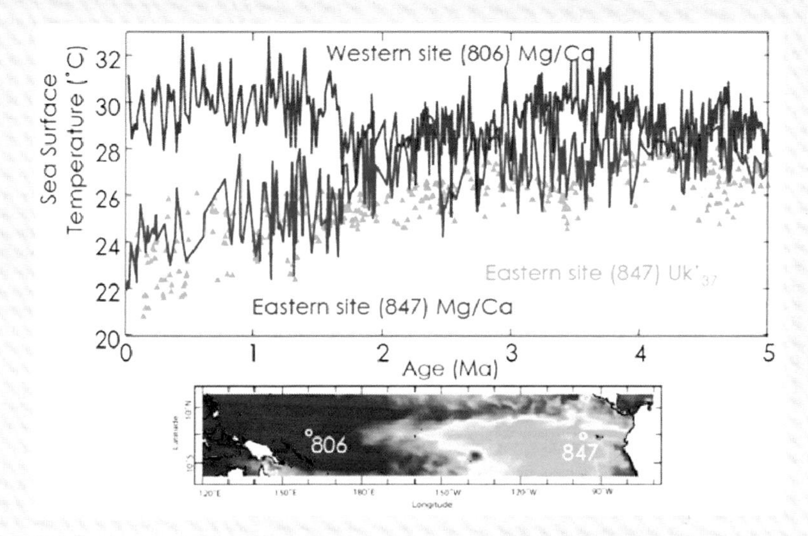

FIGURE 2.6 Changes in equatorial sea surface temperature gradients over time. The top panel shows time series of paleo-sea surface temperatures reconstructed for the western equatorial Pacific (ODP Site 806) using foraminiferal Mg/Ca ratio measurements of *Globigerinoides sacculifer*, and for the eastern equatorial Pacific (ODP Site 847) using Mg/Ca ratios of *G. sacculifer* as well as the biomarker $U^{k'}_{37}$ index. Note the reduced sea surface gradient from 5 to 3 Ma, a gradual diversion of the records after that time, and a relatively strong gradient established after about 1.7 Ma. For comparison, the lower panel shows the present-day equatorial sea surface temperature gradient and the location of the eastern (Site 847) and western (Site 806) deep-sea coring sites.
SOURCES: Compiled from Wara et al. (2005) and Dekens et al. (2008).

menting shifts in the position of climate zones and precipitation during past warm periods characterized by elevated atmospheric CO_2. For example, paleosol mineralogy and geochemistry, as well as stable isotope compositions ($\delta^{18}O$ and δD) of soil-formed minerals, fossil plants, and biogenic apatites, have been used to document substantial spatial and temporal changes in paleoprecipitation for periods spanning the past 400 million years. These proxies have documented, in particular for past warm periods, humidity and precipitation patterns substantially different from current patterns (e.g., Fricke et al., 1998; Stiles et al., 2001; Jahren and Sternberg, 2002; Sheldon et al., 2002; Tabor et al., 2002; Driese et al., 2005). Lacustrine sediments have been used to document changes in regional precipitation by using geochemical and mineralogical records to identify open flow and evaporative phases (e.g., Olsen, 1986). Quantitative assessment of relative humidity has been successfully carried out using the stable isotope compositions of Eocene cellulose (e.g., Jahren and Sternberg, 2003; Jahren et al., 2009) and leaf wax *n*-alkanes, which have a higher potential for preservation and show promise as a paleo-aridity proxy that can be used back to the Devonian (Liu and Huang, 2005; Pagani et al., 2006; Smith and Freeman, 2006).

Mass balance models of soil-formed carbonate (calcite and sphaerosiderite) using oxygen isotopes document increased precipitation rates in response to intensified hydrological cycling during the middle Cretaceous and Eocene greenhouse periods (e.g., White et al., 2001; Ufnar et al., 2002, 2004; Bowen et al., 2004; Jahren et al., 2009). For example, increased atmospheric water vapor, together with increased rainout suggested by stable isotope proxy records (Ludvigson et al., 1998), would have had the effect of enhancing the transfer of latent heat from the tropics to the high latitudes, thereby sustaining polar warmth and reinforcing greenhouse conditions (Hay and DeConto, 1999). Recent modeling studies document the additional role of high-latitude cloud feedbacks—with intensified hydrological cycling—in setting up anomalous polar warmth and maintaining depressed latitudinal temperature gradients (Abbot and Emanuel, 2007; Abbot and Tziperman 2008). Accordingly, deep-time data-model comparisons of past warm and transitional periods provide the ultimate test of science's ability to forecast the geographic patterns of hydrological response to CO_2-forced global warming and associated precipitation, evaporation, and latent heat fluxes.

SEA LEVEL AND ICE SHEET FLUCTUATIONS IN A WARMER WORLD

Rising sea level is one of the most highly visible results of a warming world and a primary concern for society (Figure 2.7). Sea level changes are dominated by changes in ocean water volume governed by the growth and decay of continental ice sheets, although other factors can contribute to smaller-magnitude fluctuations (e.g., thermal expansion of the ocean, variations in groundwater and lake storage; Miller et al., 2005). Sea level is projected to rise between ~0.4 and 1 m by the end of this century, with long-term projections of up to 7 m if the Greenland or West Antarctic ice sheets were to collapse (Alley et al., 2005; IPCC, 2007). Melting mountain glaciers could contribute another 0.5 m of sea level rise.

The present rate of sea level rise (~3 mm per year) can be expected to continue to accelerate because of positive feedbacks that act to significantly increase the warming response times and ice melting rates (Joughin et al., 2008; van de Wal et al., 2008). Currently, there is no clear consensus on the potential rate and magnitude of future ice sheet melting and sea level rise because of uncertainties in the theoretical understanding of ice sheet dynamics (Alley et al., 2005), and because strong feedbacks in response to warming (e.g., albedo, vegetation, and carbon cycling)—illustrated by the past few glacial-interglacial cycles—are not evident from historical records (Thompson and Goldstein, 2005; Carlson et al., 2008; Rohling et al., 2009). During the last interglacial (~110,000 years ago [ka])—an interval slightly warmer than today but with CO_2 at preindustrial levels—sea level was 4 to 6 m higher than at present (Rohling et al., 2008), indicating that there was deglaciation of most of Greenland (Cuffey and Marshall, 2000) and probably also a contribution from Antarctica. Notably, rates of sea level change during glacial terminations and interglacial periods in the past few 100,000 years were substantially faster, between >1 and ~5 cm per year (Carlson et al., 2008; Rohling et al., 2009) than recent rates of rise (~3 mm per year).

Moreover, the deep-time sedimentary record contains evidence of past sea level changes at rates an order of magnitude or more higher than the current observed rate of rise (Miller et al., 2005). Because reconstructions of paleoeustasy are typically based on passive margin stratigraphic records or indirect geochemical records such as the oxygen isotope composition of marine sediments and microfauna, such estimates come with uncertainties. However, because positive feedbacks to warming are explicitly not considered in current projections (e.g., IPCC, 2007) and sea level can change at rates and magnitudes greater than typically considered, there is an obvious need to better understand how passive margin successions record past glacioeustasic sea level changes and their relationship to variations in atmospheric pCO_2 and surface temperatures.

Despite such uncertainty, the geological record is the only repository that can place constraints on the sensitivity of ice sheets and *equilibrium* sea level to rapid (millennia or less) climate change during past warmings that led to collapse of ice sheets of the scale of, or larger than, the Greenland or Antarctic ice sheets. During the Middle Pliocene warming (3.5-3.0 Ma),

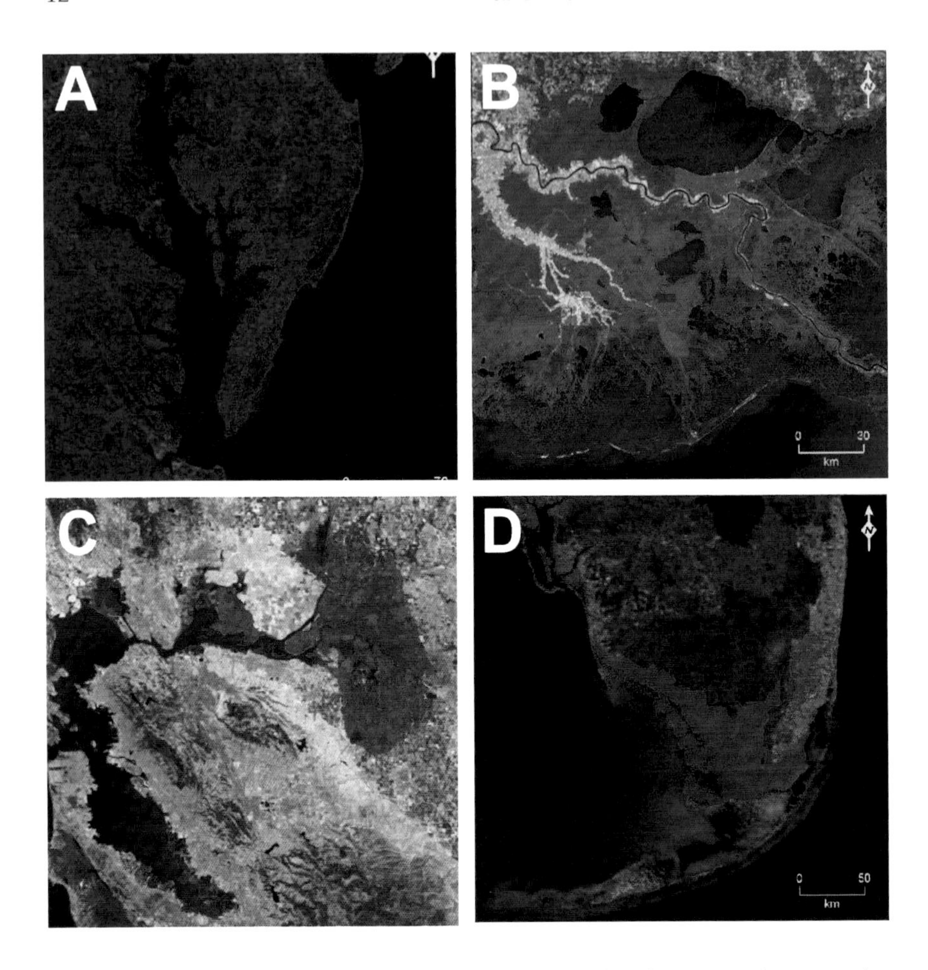

FIGURE 2.7 Effects of a 1-m rise in sea level (inundated areas with connectivity to the sea shown in red; see Weiss et al., 2011)—in the absence of human intercession—for four areas of the U.S. coastline: (A) Chesapeake Bay area; (B) New Orleans area; (C) San Francisco Bay area, and (D) South Florida. Because of the time needed for atmospheric CO_2 and surface warming to reach equilibrium, this amount of sea level rise might be anticipated even if anthropogenic sources of CO_2 were to cease today. In addition to local effects, inundation of the San Francisco Bay area and south Florida would have substantial impacts on water resources in both California and Florida.
SOURCE: Images courtesy Jeremy Weiss and Jonathan Overpeck, University of Arizona.

when CO_2 levels were slightly higher than at present (330 to 415 parts per million by volume; Pagani et al., 2010) and comparable to projections for the coming decade, sea level is estimated to have been 15-25 ± 5 m, and possibly up to 36 m, above the present level (Wardlaw and Quinn, 1991; Shackleton et al., 1995; Naish et al., 2009). It is clear that equilibrium sea level in response to current warming may be substantially higher than projected by IPCC models given that they do not include dynamic processes that were operative in past warm periods (Rohling et al., 2009). Therefore, even with stabilization of atmospheric CO_2 at current levels, it is likely that sea level will continue to rise over the next few centuries at much faster rates—up to three to four times faster than at present—than has hitherto been considered by long-term projections.

Evidence for continental ice sheets during past long-lived greenhouse periods defines a "climate-glaciation paradox" which in turn provides an opportunity to uniquely address two issues that cannot be tested in studies of the more recent glacial-interglacial fluctuations of the Cenozoic. First, the geological records of glacial events during past greenhouse warm periods illustrate ice sheet stability and long-term equilibrium sea level change during times of substantially elevated atmospheric pCO_2, major climate perturbations, and in many cases, complete deglaciation. For example, geological records offer sedimentological (Alley and Frakes, 2003), stratigraphic (Gale et al., 2002, 2008; Miller et al., 2003, 2004; Plint and Kreitner, 2007), and geochemical (Stoll and Schrag, 1996, 2000; Bornemann et al., 2008) evidence—albeit indirect—for moderate to high-magnitude (<25 m) sea level fluctuations and, if ice-driven, the possible existence of continental ice during the protracted greenhouse periods of the Cretaceous and early Eocene. The development of ephemeral (<200,000 years) and small to moderate ice sheets in Antarctica (Figure 2.8; Miller et al., 2005; Miller, 2009) documents the climate mechanisms and feedbacks that permitted the repeated development of glaciations during otherwise hothouse climates (Bornemann et al., 2008). We can also use the geological record to investigate the dynamics of ice-driven sea level change and deglaciation from such records of ice during times of anomalously warm tropical oceans (up to 35°C) and forested poles.

Major variations in the expansion and contraction of ice sheets throughout the Late Paleozoic Ice Age (Fielding et al., 2008; Rygel et al., 2008; Bishop et al., 2009) may have involved expansion of continental ice into the tropics (Soreghan et al., 2008), followed by protracted periods of minimal glaciation that were likely CO_2 forced (DiMichele et al., 2009). Data-model comparisons for this ice age (Poulsen et al., 2007a; Peyser and Poulsen, 2008; Horton et al., 2010) indicate the influence of high-latitude vegetation-climate feedbacks and the need for pCO_2 changes of much greater magnitude than those associated with more recent interglacial-

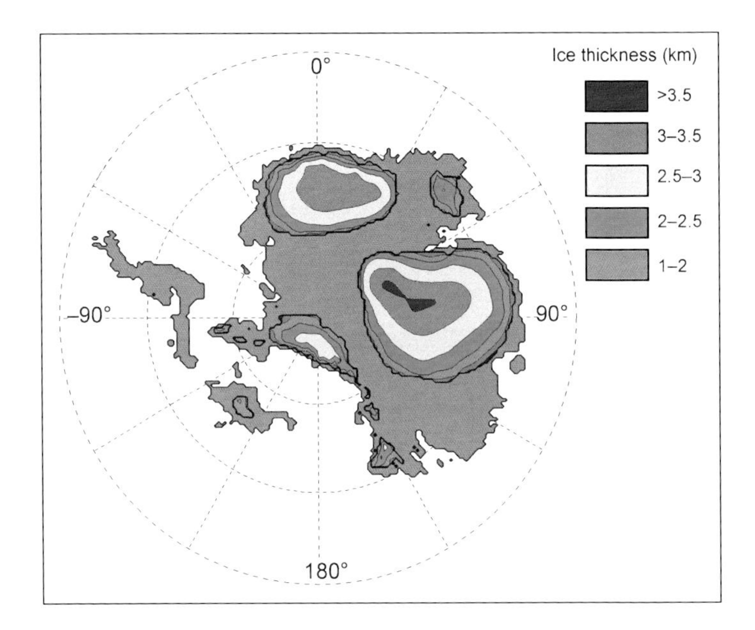

FIGURE 2.8 Modeled distribution of greenhouse ice sheets on Antarctica. Relatively small (~25 m) fluctuations in sea level have been associated with Antarctic ice (Galeotti et al., 2009), which would likely have built up in areas of high topography or near coastal moisture.
SOURCE: DeConto and Pollard (2003), reprinted by permission of Macmillan Publishers Ltd.

glacial cycles, and highlight the large-scale reorganization of atmospheric circulation and regional precipitation patterns between periods of extensive glaciation and the warmer "interglacial" periods with greatly diminished continental ice. This late Paleozoic icehouse, which is calibrated to the orbital timescale (Strasser et al., 2006; Davydov et al., 2010), terminated with the full collapse of expansive southern hemisphere ice sheets, thereby serving as the only "vegetated" analogue for an icehouse-to-greenhouse transition (Montañez et al., 2007).

The second issue that can be assessed uniquely in deep-time records is an improved understanding of the origins of short-term (10,000 years to <2 million years) sea level fluctuations. Some of these sea level changes could reflect the development of small ice sheets, where direct evidence for ice from glacial sediments or oxygen isotope excursions is often lacking or ambiguous (Miller et al., 2005; Haq and Schutter, 2008). Rapid sea level changes during past warm periods also raise the question of whether ephemeral ice sheets were common to all past warm periods.

Short glacial advances are known to be associated with rare alignments of the orbital obliquity and eccentricity cycles (Zachos et al., 2001a; Pälike et al., 2006a,b), suggesting that orbital configuration may predispose the climate system to glaciation and produce short sea level falls. It is also possible that there are nonglacioeustatic—and perhaps nonclimatic—causal mechanisms that could influence ice sheet dynamics and sea level response with continued warming.

EXPANSION OF OCEANIC HYPOXIA IN A WARMER WORLD

One forecast consequence of global warming is the widespread expansion of oceanic hypoxia. Projections anticipate a decline in ocean O_2 by as much as 30 percent in the next several centuries, with the potential to expand the area of hypoxic seas from the current ~9.1 percent to as much as 61 percent of the total ocean (Shaffer et al., 2009; Keeling et al., 2010). Excess nutrient loading from river runoff, in combination with warming waters and increasing pCO_2, presently results in annual or permanent expanses of hypoxia or anoxia on continental shelves and estuaries—well-publicized examples are Chesapeake Bay (Adelson et al., 2001) and the northern Gulf of Mexico (Malakoff, 1998). Globally, more than 400 "oceanic dead zones"—so named because of their inability to support marine animal life (Diaz and Rosenberg, 2008; Rabalais et al., 2009)—have been identified, most of which have appeared within the last 50 years and now cover a cumulative area of ~245,000 km^2 (Malakoff, 1998; Diaz and Rosenberg, 2008). Models suggest that hypoxic and anoxic zones will continue to increase in response to warming and increasing pCO_2, making this one of the important risks associated with current warming (Stramma et al., 2008; Shaffer et al., 2009). The extent and rate at which they will increase in the coming century, however, are unknown.

Periods of anoxia in the ancient record provide models for understanding the biological and environmental consequences of widespread hypoxia (Gooday et al., 2009). In each past case, the combined effects of increased surface ocean temperatures and reduced vigor of overturning circulation—effects anticipated with the current global warming—led to dramatically lower levels of dissolved oxygen throughout the water column (Broecker, 1999). In the geological past, episodes of widespread marine hypoxia have been associated with biotic crises of aerobic marine organisms (Sepkoski, 1996), such as the Middle-Late Devonian crises, the Permian-Triassic extinction—the largest mass extinction of the past half billion years, and repeated oceanic anoxic events throughout the mid to late Cretaceous. These ancient examples provide an invaluable archive of triggers, thresholds, rates of onset and recovery, and the spatial distribution of oxygen deficiency in the oceans.

Inferred widespread oxygen depletion in latest Permian to earliest Triassic oceans (Wignall, 2007) provides an example of oceanic hypoxia that developed in multiple ocean basins. Oxygen deficiency in these paleo-oceans developed to a degree sufficient for free hydrogen sulfide to accumulate in the upper mixed layer of the surface ocean (Kump et al., 2005; Meyer et al., 2008). Although the intensity of oxygen depletion varied regionally, evidence for hydrogen sulfide, a compound that is toxic to most marine organisms at even relatively low concentrations (~100 μmol), has been found in all contemporary Permo-Triassic oceans (the Tethys and Panthalassa; Isozaki, 1997; Grice et al., 2005; Algeo et al., 2008). Although the exact cause of this nearly global anoxic episode remains under debate, there is little doubt that greenhouse gas-induced global warming played an important role (Box 2.4; Korte et al., 2005; Kearsey et al., 2008). Warming of deep-water source regions (Hotinski et al., 2001) potentially reduced the vigor of oceanic circulation (Isozaki, 1997; Kiehl and Shields, 2005) and elevated the nutrient flux to the oceans in response to increasing mean temperatures. In turn, this may have provided the trigger for an oceanic anoxic event of this enormity (Winguth and Maier-Reimer, 2005; Meyer et al., 2008; Algeo et al., 2010).

Mid-to-late Mesozoic marine successions document repeated expansions of anoxia, with the majority occurring during protracted periods of "supergreenhouse" conditions (Jenkyns, 1988; Sageman et al., 2006; Hesselbo et al., 2007). These events were associated with the turnover of oceanic biota, including major evolutionary bursts in planktonic foraminifera and the extinction of several nannoplankton groups (Leckie et al., 2002). These geological-scale "anoxic events" (Box 2.5)—with their attendant changes in surface temperatures and continental weathering (i.e., phosphorus fluxes to the ocean)—have been linked to increased levels of atmospheric greenhouse gases brought on relatively slowly through volcanism or by more rapid release following magmatic intrusion into organic-rich sediments (Tejada et al., 2002; Svensen et al., 2004; Turgeon and Creaser, 2008; Barclay et al., 2010). The increased weathering led to substantially increased nutrient fluxes to the oceans (Hochuli et al., 1999; Weissert and Erba, 2004; Tsandev and Slomp, 2009; Adams et al., 2010), analogous to the present-day flux of agricultural fertilizers to the oceans (Rabalais et al., 2009).

The dynamics of ancient anoxic intervals are particularly well studied in the Mediterranean basin. There, the hypoxic sediment record began about 14 million years ago (Mourik et al., 2010) and became more pronounced in the Plio-Pleistocene (Figure 2.11), particularly during the past 2 million years (Emeis and Weissert, 2009). Hypoxia is associated with organic-rich sediments (up to 30 weight percent organic carbon) as well as evidence for surface ocean warming, freshwater inputs that shut down basin overturning, and the regeneration of nutrients into the surface

BOX 2.4
A Literal Smoking Gun for the End-Permian Extinction

In the minds of most scientists, the case of what killed the dinosaurs is closed. The culprit was an impact by a 10-km-wide asteroid, with the consequent heat and chemical modifications of the atmosphere causing environmental disturbance that affected both land and marine flora and fauna and in some cases resulted in extinctions. In contrast, what caused the largest of all mass extinctions—at the end of the Permian at 252 Ma—is a "cold case" that has only recently received new information that might ultimately lead to the "killer." A key piece of evidence is the remarkable correspondence in ages (Figure 2.9) between this event and vast eruptions of the Siberian Traps volcanoes, one of the largest volcanic eruptions in Earth history (Reichow et al., 2009). Volatiles released from the magma may themselves have created a sizable environmental disturbance, but the key to the deadliness of the Siberian Traps eruptions seems to be magma penetration through thick sequences of limestone, evaporite, and coal (Svensen et al., 2009), baking these materials and releasing immense quantities of sulfur dioxide and gaseous halides that cooled the climate and disrupted the ozone layer on short timescales (Beerling et al., 2007), yet warmed it on longer (multimillennial) timescales by the release of large amounts of CO_2 that boosted atmospheric CO_2 levels.

The abrupt and prolonged global warming would have been stressful to biota by itself, but for marine organisms the resulting reduction in oxygen solubility and reduced oxygen delivery to the deep ocean led to widespread seafloor anoxia and permitted the accumulation of toxic substances, including hydrogen sulfide, in the deep ocean (Meyer et al., 2008). Upwelling of hydrogen sulfide into surface waters poisoned oxygen-dependent organisms and encouraged sulfur-metabolizing organisms to thrive. Environmental recovery apparently was quite slow—carbonate rocks with anomalous marine cements and microbial reefs (Woods et al., 2007) that are similar to those of the Proterozoic (Grotzinger and Knoll, 1995) characterize the first few million years of the Triassic, as do depauperate faunas of small bivalves. Full recovery did not occur for several million years, with the composition of subsequent Mesozoic communities reflecting the complex dynamics of recovery and diversification and large perturbations in global biogeochemical cycling (Payne et al., 2004). The nature of the biotic recovery, however, remains poorly understood and will require interdisciplinary studies of physicochemical and biological processes integrated with quantitative models of diversification dynamics and global climate and biogeochemistry.

continued

BOX 2.4 Continued

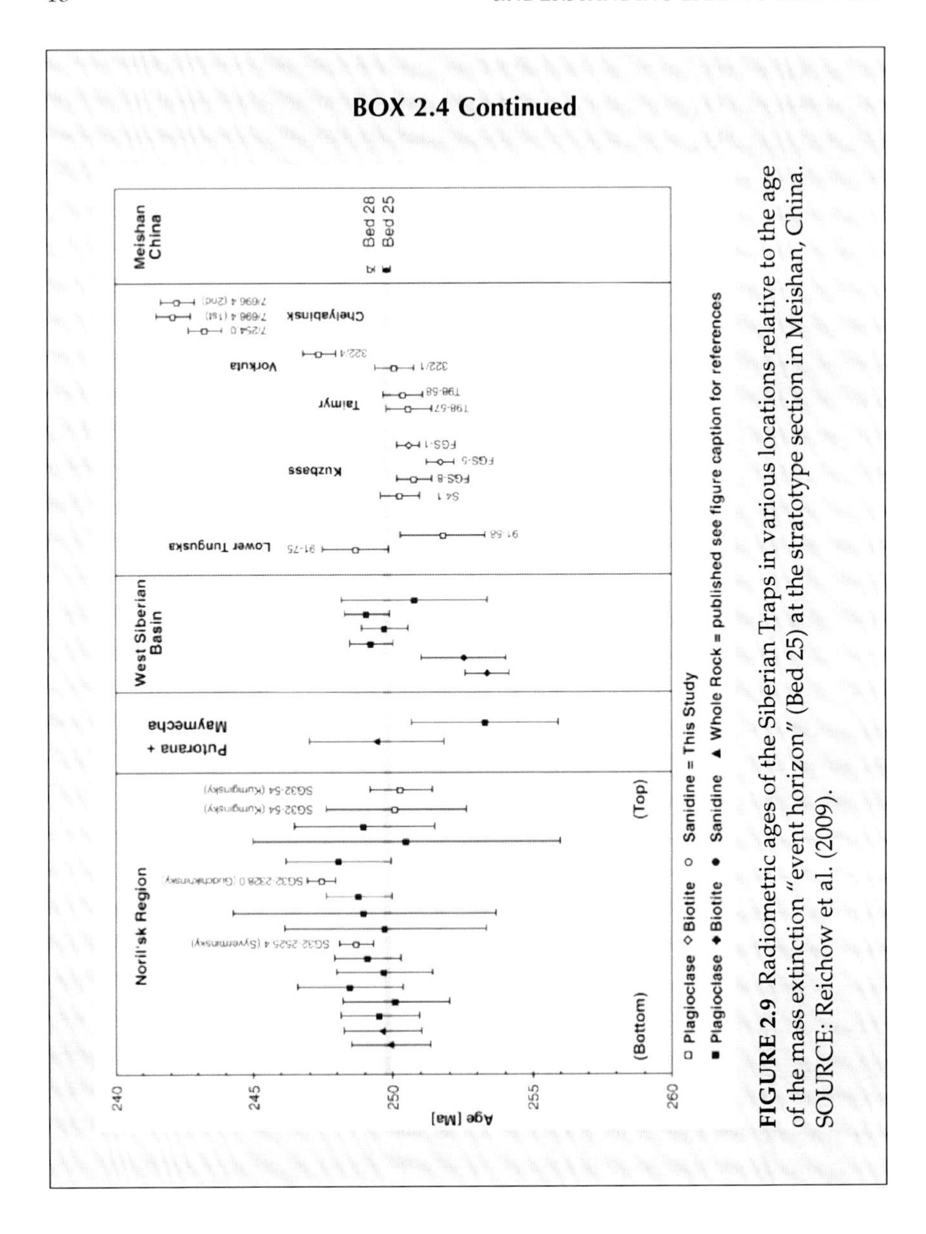

FIGURE 2.9 Radiometric ages of the Siberian Traps in various locations relative to the age of the mass extinction "event horizon" (Bed 25) at the stratotype section in Meishan, China. SOURCE: Reichow et al. (2009).

BOX 2.5
Triggering of Oceanic Anoxic Events

Widespread, recurrent intervals of oxygen-deficient conditions in the water column (referred to as oceanic anoxic events, or OAEs) were a common phenomenon during the Mesozoic (Arthur et al., 1990; Hochuli et al., 1999; Hesselbo et al., 2007). Positive $\delta^{13}C$ isotope excursions persisting over tens to hundreds of thousands of years are characteristic of most of these organic-rich OAEs, interpreted as resulting from increased burial of ^{12}C-enriched sediments (Scholle and Arthur, 1980; Weissert, 1989; Menegatti et al., 1998). Some of these anoxic events (notably OAE 1b and 1d) appear to be ancient analogues of the well-studied Mediterranean sapropels that were produced by the hydrological suppression of ocean overturning in a marginal silled basin (Erbacher et al., 2001; Wilson and Norris, 2001). Other events (e.g., OAE 1a and 2) were global events whose origin is not clearly tied to tectonic preconditions. Still other brief periods of widespread anoxia may have an origin in greenhouse gas-driven warming. Indeed, the more recent discovery of large-magnitude but much shorter-lived (10^3 to 10^4 years) negative isotope excursions at the onset of several of these Mesozoic OAEs is compelling evidence for greenhouse gas forcing of these abrupt climate events, possibly by methane release from seafloor gas hydrates (Menegatti et al., 1998; Hesselbo et al., 2000; Jahren et al., 2001), methane release by magmatic intrusion into organic-rich sediments (e.g., coals; Svensen et al., 2004; McElwain et al., 2005), or other greenhouse gas sources such as volcanism (Svensen et al., 2007; Hermoso et al., 2009).

Nested within the longer-term positive carbon isotope excursion, some Cretaceous black shales show a striking cyclicity over submeter thicknesses, which is interpreted to record oceanic hypoxia responding to minor climate effects caused by orbitally driven changes in atmospheric circulation. For example, mid-Cretaceous black shales from paleotropical West Africa at ODP Site 959 provide evidence in their total organic carbon content (2-12 percent), clay mineralogy, biomarkers indicative of euxinia, and elemental ratios for coupled, centennial timescale changes in atmospheric circulation and marine circulation coincident with strengthening of the African monsoon (Wagner et al., 2004; Beckmann et al., 2005) (Figure 2.10). The rapid response and strong variability recorded by these Cretaceous OAE intervals demonstrates the sensitivity of oceanic conditions to perturbation of atmospheric circulation and continental weathering brought on by global warming.

Can the deep-time record tell us more about an inherent instability of climate over short timescales (decadal to centennial) if it is analyzed at appropriate resolution? These examples would suggest so and emphasize the enormous potential for unlocking the lessons about warm-Earth climate states that are recorded solely in the deep-time geological archive.

continued

BOX 2.5 Continued

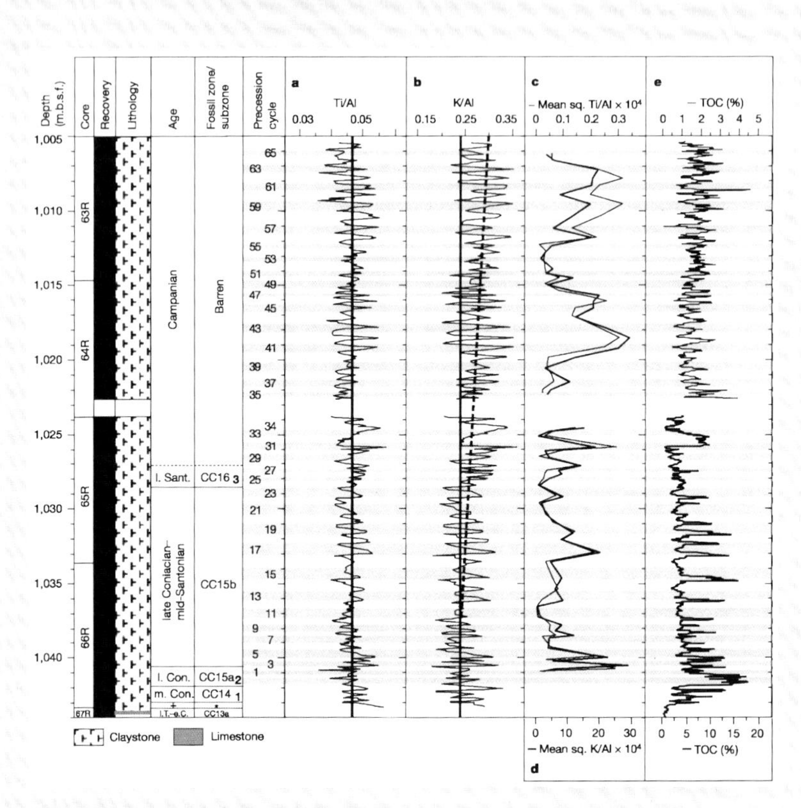

FIGURE 2.10 High-resolution plots showing multiple precessional cycles during Cretaceous OAE 3 (~85 Ma) from Ocean Drilling Program Site 959 in the eastern equatorial Atlantic. There is abrupt and high-frequency variation in total organic carbon, with higher values indicating anoxia. These cycles have been interpreted as resulting from changes in the Cretaceous intertropical convergence zone that caused variations in river discharge from tropical western Africa.
SOURCE: Beckmann et al. (2005).

waters that trigger massive blooms of bacteria and diatoms (Bianchi et al., 2006; Capozzi et al., 2006; Rohling et al., 2006; Gallego-Torres et al., 2007; Emeis and Weissert, 2009). The Mediterranean record appears to have parallels with some Cretaceous events, particularly those developed in the narrow proto-North Atlantic (Erbacher et al., 2001).

BIOTIC RESPONSE TO A WARMER WORLD

Although Earth's biota contributes to, and is affected by, a variety of important climatic feedbacks that are incompletely understood in terms of process and magnitude, there is little doubt within the scientific community that anthropogenic climate change is having a large and sustained impact on Earth's ecosystems (Parmesan and Yohe, 2003; Rosenzweig et al., 2008). Changes in the distribution and ecological composition of vegetative land cover (e.g., replacement of evergreen forests by deciduous forests) can lead to changes in mean global and regional albedos, altering both the total amount of incoming solar radiation absorbed at the Earth's surface and its spatial distribution (Baldocchi et al., 2000; Marland et al., 2003). Ecosystem changes have already accompanied the rapid loss of snowfields and sea ice, with consequent decrease in surface albedo in high-latitude regions, as a result of the global warming of the past century (ACIA, 2004; Chapin et al., 2005). Changes in terrestrial vegetation also lead to changes in evapotranspiration and soil moisture content and, in turn, cloud cover and water vapor mass in the atmosphere (Hennessy et al., 1997; Alpert et al., 2006). In the marine environment, changes in upwelling and nutrient availability have the potential to influence the rate of dimethyl sulfide production by phytoplankton, thereby altering the concentration of cloud condensation nuclei and changing the albedo and other optical properties of stratus clouds over oceans (Boucher and Lohmann, 1995; Schult et al., 1997; Kump and Pollard, 2008). Vegetation-climate feedbacks, however, involve complex, nonlinear interactions with competing effects, resulting in an uncertain net response to climatic forcing. Nevertheless, general circulation models incorporating vegetation-climate feedbacks generally yield higher climate sensitivities (up to 5.5°C per CO_2 doubling; Cox et al., 2000) relative to models lacking such feedbacks, reflecting the direct influence that vegetation has on radiation fluxes, water cycling, and latent heat transport on Earth's surface.

The Pleistocene paleoclimate record provides well-documented examples of the impact on, and influence of, vegetation on climate shifts associated with the gentle glacial-interglacial oscillations in atmospheric CO_2 over the past few hundred thousand years (Peteet and Mann, 1994; Kneller and Peteet, 1999; Gillespie et al., 2004). The relative importance of such feedbacks, however, is likely to change as atmospheric CO_2 concentrations

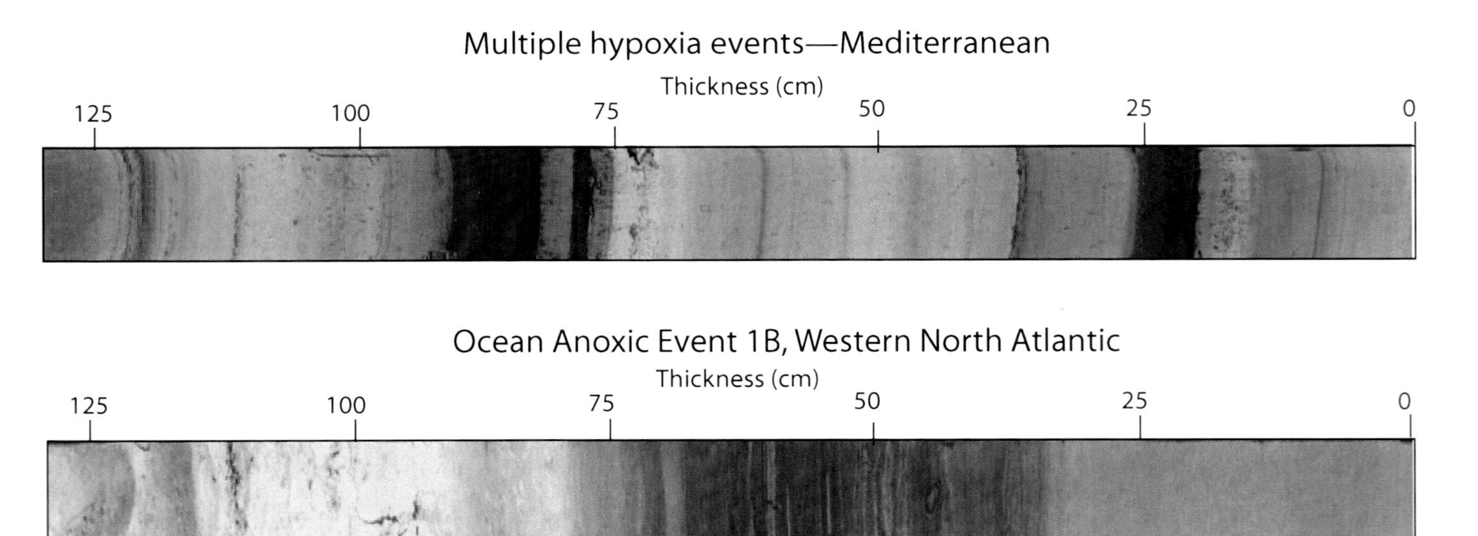

FIGURE 2.11 Examples of ancient hypoxic episodes in a Plio-Pleistocene drill core from the Mediterranean. Black bands in the upper photo (from Ocean Drilling Program Site 964) are "sapropels"—layers rich in organic carbon—formed when the surface waters of the Mediterranean abruptly warmed, became fresher, and ceased to circulate as they do today. Warming and high nutrient supply led to blooms of algae and bacteria, preserved as a layer of organic carbon on the seafloor. Lower photo shows a core from the western North Atlantic (from ODP Site 1049) in which a similar layer rich in organic carbon was deposited during an event 112 million years ago that was broadly analogous to the Mediterranean hypoxic events.
SOURCE: Images courtesy Integrated Ocean Drilling Program Science Services.

continue to increase well above the levels of the Pleistocene interglacials and as the geographic distributions of climate zones change. For example, higher CO_2 levels are expected to saturate the CO_2 fertilization effect, resulting in a shift of the terrestrial biosphere from a net sink to a net source of carbon sometime within this century (Cao and Woodward, 1998; Cox et al., 2000). Furthermore, as the surface oceans warm and become less alkaline with increasing atmospheric CO_2, carbonate-bearing animals will be strongly impacted (e.g., see Box 2.6), further perturbing biota-climate feedbacks compared with those reconstructed from the recent past.

The deep-time geological record, in particular the record of warm periods of higher atmospheric pCO_2 and including the transitions into and out of these periods, has the potential to yield unique insights into the nature and rate of biotic response to climate perturbation as well as into the biota-climate feedbacks accompanying global warming. For example, the mid-Paleozoic "greening" of continents, marked by the evolution and spread of vascular land plants (Gensel and Andrews, 1987; Beerbower et al., 1992), records a large-scale natural experiment in the climatic effects of vegetation—reflecting the contrast between a largely unvegetated pre-Devonian world compared with one that was heavily vegetated—that has been linked to major changes in atmospheric CO_2 and a vastly different hydrological regime (Algeo et al., 1995, 2001). Another example of the potential of the deep-time record is provided by the repeated major restructuring and turnover within terrestrial floral communities that occurred in step with recurrent shifts in surface temperature, precipitation levels, seasonality, and soil moisture during the demise of the Late Paleozoic Ice Age at ~295-260 Ma, the vegetated Earth's only analogue of a CO_2-forced icehouse-to-greenhouse transition (see Box 2.7).

More recently, the gradual but extreme warming (perhaps up to >30 to 42°C in the tropics) of the early Eocene greenhouse (Box 2.8) may have triggered a major tropical vegetation die-off, with substantial changes in evapotranspiration fluxes, precipitation, albedo, surface temperature, and carbon feedbacks (Huber, 2008). During the transient global warming and short-term aridity of the Paleocene-Eocene Thermal Maximum (PETM) major restructuring among terrestrial biomes resulted in expansion in the latitudinal range of subtropical and tropical rainforests (Wing et al., 2005). Oxidation of the terrestrial biosphere at the Paleocene-Eocene boundary may have released several gigatons of carbon into the atmosphere, substantially amplifying the existing greenhouse warming and its climate effects (Kurtz et al., 2003).

The potential vulnerability of modern biotic communities to catastrophic disruption (Jackson et al., 2001; Chase and Leibold, 2003) is an issue designated as one of the "grand challenges" in the environmental sciences (NRC, 2001). Globally, current extinction rates are estimated to

BOX 2.6
Impact of Past and Future Climate Change on Coral Reefs

Healthy coral reef ecosystems develop under a relatively narrow range of ocean temperatures and chemistry (Kleypas et al., 1999) and are therefore sensitive indicators of environmental conditions. Global change models predict that reef systems, with their abundant biodiversity, will be exposed to higher ocean temperatures and increasingly more acidic waters in the next century (Hoegh-Guldberg et al., 2007; see Figure 2.12). Indeed, research suggests that global climate change has already caused steep declines in coral growth on reef systems around the world (Hoegh-Guldberg, 1999). Culturing experiments with corals in acidified waters show that skeleton growth drops as acidity increases and, in extreme cases, coral colonies can lose their skeletons completely and grow as soft-bodied anemone-like animals (Fine and Tchenov, 2007). In fact, ocean acidification may vie with global warming as the most severe threat to marine ecosystems (Hoegh-Guldberg et al., 2007; De'ath et al., 2009). Reef systems, however, are intrinsically complex structurally and ecologically, making it difficult to evaluate the likely impact of future global change on modern reefs based solely on studies of present-day systems. The geologic record of fossil reef evolution provides opportunities to study the response of reef ecosystems to past episodes of increased global temperatures and ocean acidification.

The coral reef crisis occurring in modern oceans may be the sixth such major reef crisis recorded in the past 500 million years of marine metazoan evolution. Four of the previous five metazoan reef crises appear to have been driven by greenhouse gas-forced global warming that was probably associated with ocean acidification (Veron, 2008; Kiessling and Simpson, 2010). At least three of these reef crises were associated with massive release of greenhouse gases into the oceans and atmosphere, leading to pCO_2 increases analogous to—or perhaps even greater than—those anticipated for Earth's future. For example, major reef crises during the Early Jurassic and during the Cretaceous were associated with massive releases of volcanic CO_2 to the atmosphere that led to global warming, oceanic anoxia, and quite likely ocean acidification (Knoll et al., 1996; Svensen et al., 2007; Hermoso et al., 2009). One of the major reef crises occurred at the same time as the best-documented case of greenhouse gas-induced ocean acidification in the geological record, the Paleocene-Eocene Thermal Maximum (PETM) of 56 Ma (described in more detail in the next chapter). Although coral-algal reefs began to decline throughout the Tethyan region in the early Eocene due to the development of very warm (~30-35°C) tropical sea surface temperatures (Scheibner and Speijer, 2008) (Figure 2.13), PETM extinction rates indicate that ocean acidification must have been a major

continued

FIGURE 2.12 Extant examples of reefs from the Great Barrier Reef that are used as analogues for the ecological structures anticipated for atmospheric CO_2 values of (A) 380 ppmv, (B) 450-500 ppmv, and (C) >500 ppmv (Hoegh-Guldberg et al., 2007). Scenario C corresponds to a +2°C increase in sea temperature. The atmospheric CO_2 and temperature increases shown are those for the scenarios and do not refer to the particular locations photographed.
SOURCE: Photographs by and with permission of Ove Hoegh-Guldberg, Global Change Institute, University of Queensland.

BOX 2.6 Continued

causative factor along with global warming (Kiessling and Simpson, 2010). Notably, the deep-time record of this major reef crisis uniquely captures the consequences on larger-scale marine ecosystems that might be anticipated with the future loss of reefs.

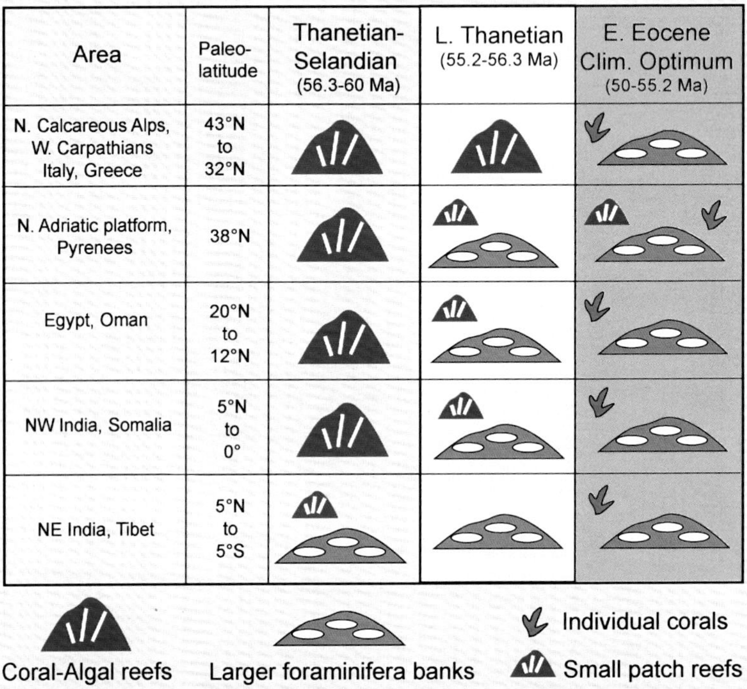

Area	Paleo-latitude	Thanetian-Selandian (56.3-60 Ma)	L. Thanetian (55.2-56.3 Ma)	E. Eocene Clim. Optimum (50-55.2 Ma)
N. Calcareous Alps, W. Carpathians Italy, Greece	43°N to 32°N			
N. Adriatic platform, Pyrenees	38°N			
Egypt, Oman	20°N to 12°N			
NW India, Somalia	5°N to 0°			
NE India, Tibet	5°N to 5°S			

Coral-Algal reefs

Larger foraminifera banks

Individual corals

Small patch reefs

FIGURE 2.13 Paleogene reef history from the Mediterranean area and southern Asia. Coral-algal reefs that are widespread in the early Paleogene largely disappeared during the peak of greenhouse warming in the Early Eocene Climatic Optimum, and were replaced by carbonate mounds formed by larger benthic foraminifera (nummulite banks). SOURCE: Modified from Scheibner and Speijer (2008).

BOX 2.7
Climate-Driven Restructuring of Late Paleozoic Tropical Forests

Integration of climate proxy records with tropical paleobotanical archives from the Late Paleozoic shows repeated climate-driven ecosystem restructuring of paleotropical flora in step with climate and pCO_2 shifts, illustrating the biotic impact associated with past CO_2-forced turnover to a permanent ice-free world (Montañez et al., 2007; DiMichele et al., 2009). Wetland flora—consisting of ferns and pteridosperms, sphenopsids, and lycopsids—was rapidly replaced in the earliest Permian by dryland flora that diversified in the now seasonally dry habitats created by an abrupt shift from ever-wet to semiarid conditions. Tree and fern-rich floras reappeared during wetter, cooler conditions of the subsequent glaciation at ~285 Ma characterized by lowered pCO_2 (Figure 2.14). Such dramatic floristic changes occurred with each climate transition during the final stage of the Late Paleozoic Ice Age.

The fact that these temporally successive floras tracked climatic conditions and contained progressively more evolutionarily advanced lineages suggests that evolutionary innovation occurred in extrabasinal areas and was revealed by climate-driven floral migration into lowland basins. One such scenario occurred during the return to cold conditions at the close of the Early Permian when unique seed-plant assemblages, not observed again until the Late Permian (conifers) and Mesozoic (cycads), migrated into lowland basins. Climate transitions drove macroevolution in the oceans as well, including significant changes in marine invertebrate biodiversity coincident with the appearance of a diverse array of early terrestrial vertebrate lineages and major restructuring of floral biomes (Clapham and James, 2007).

continued

BOX 2.7 Continued

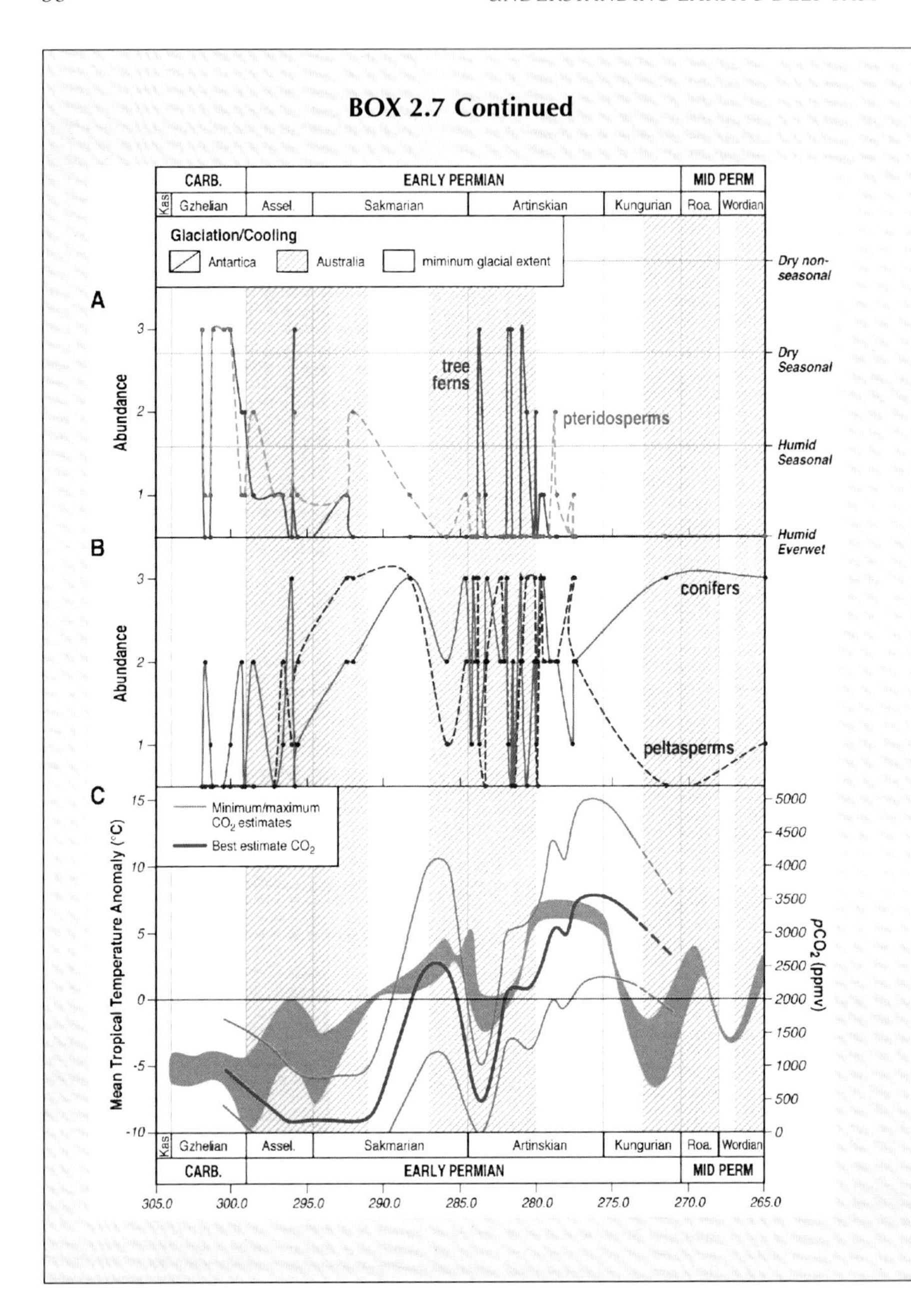

FIGURE 2.14 Floral abundance patterns (A and B) in the paleotropics during the latest Pennsylvanian through Middle Permian, plotted against (C) estimated pCO_2 (blue line) and paleo-sea surface temperatures (red band). Periods of glaciation or widespread cooling in the high southern latitudes are shown by blue bars. Top panel (A) shows the temporal distribution of typical latest Carboniferous wetland floras (ferns and pteridosperms, sphenopsids, and lycopsids). The middle panel (B) illustrates the temporal distribution of dryland floras (conifers, callipterids and other seed plants) that diversified in seasonally dry habitats. The short-term intercalation of the two floras—at the likely millennial scale—occurred with the return of wetland floras in the mid Early Permian transient glaciation under cooler and wetter conditions brought on by significantly lowered pCO_2 and renewed glaciation.
SOURCE: Modified after Montañez et al. (2007).

BOX 2.8
Biome Distributions in the Cretaceous-Early Eocene Hothouse

The discovery more than a century ago of coal seams, fossil forests, and fossil leaves of warm temperate trees above the Arctic Circle on the west coast of Greenland, more than 1,000 km north of the modern tree line, was an early indicator of anomalous warmth at high latitudes in the past. That arctic regions more than 50 Ma had been forested as far north as there was land, despite a continental configuration similar to that of the present day, became increasingly more apparent with the discovery of hundreds of similar sites on arctic islands and across arctic Asia and North America (Spicer et al., 2008). Northern hemisphere Cretaceous-early Eocene polar forests were fundamentally different from present-day boreal forests, which do not grow north of the Arctic Circle, as they are dominated by deciduous conifers related to the bald cypress and dawn redwood and by a variety of deciduous broadleaf trees. Leaf margin analysis of Paleocene-Eocene floras from arctic Canada on Axel Heiberg Island (at 78°N) yield mean annual temperatures of 10° ± 2°C (Basinger et al., 1994), in striking contrast to the modern mean annual temperatures of minus 30°C. The fossil floral record indicates that in this warmer world, both subtropical and tropical rainforests had greatly expanded latitudinal ranges (Figure 2.15).

Subtropical conditions in the polar Arctic are further indicated by the occurrence of early Eocene crocodiles, turtles, and snakes on Ellesmere Island at 80°N at this time (Dawson et al., 1976; Markwick, 2007). Subsequent discovery of fossil mammals and plants, related to contemporary biotas in France and Wyoming, confirmed the hypothesis that Arctic Canada at this time was part of a warm temperate land connection between Europe and North America (Hickey et al., 1983). The recent and surprising discovery of the aquatic fern *Azolla* in 47 Ma Eocene sediments in an ACEX Integrated Ocean Drilling Program (IODP) core in the middle of the Arctic Ocean (Brinkhuis et al., 2006) adds an almost surreal element to this vignette of crocodile-infested subtropical swamp forests on the shores of a warm, fresh arctic ocean covered with floating aquatic plants. Studies of such ice-free, high-latitude, deep-time analogues are important scientific windows into how the Arctic ecosystem might operate in the absence of permanent sea ice or in fully deglaciated conditions.

In a world with forested poles and tropical midlatitudes, the nature of the equatorial realm is a serious question. Recent estimates of sea surface temperatures for the Late Cretaceous to Eocene tropics, based on well-preserved marine microfossils, suggest that temperatures may have exceeded 35-40°C (Huber, 2002; Norris et al., 2002; Pearson et al., 2007)—the absence of equatorial coral reefs may have been because seawater was too hot. Ample evidence from equator to pole shows that the last greenhouse was a very different place from today, and that the composition and distribution of biomes were wholly different from the present—it was not just a warmer world, but rather a completely different world from the present day.

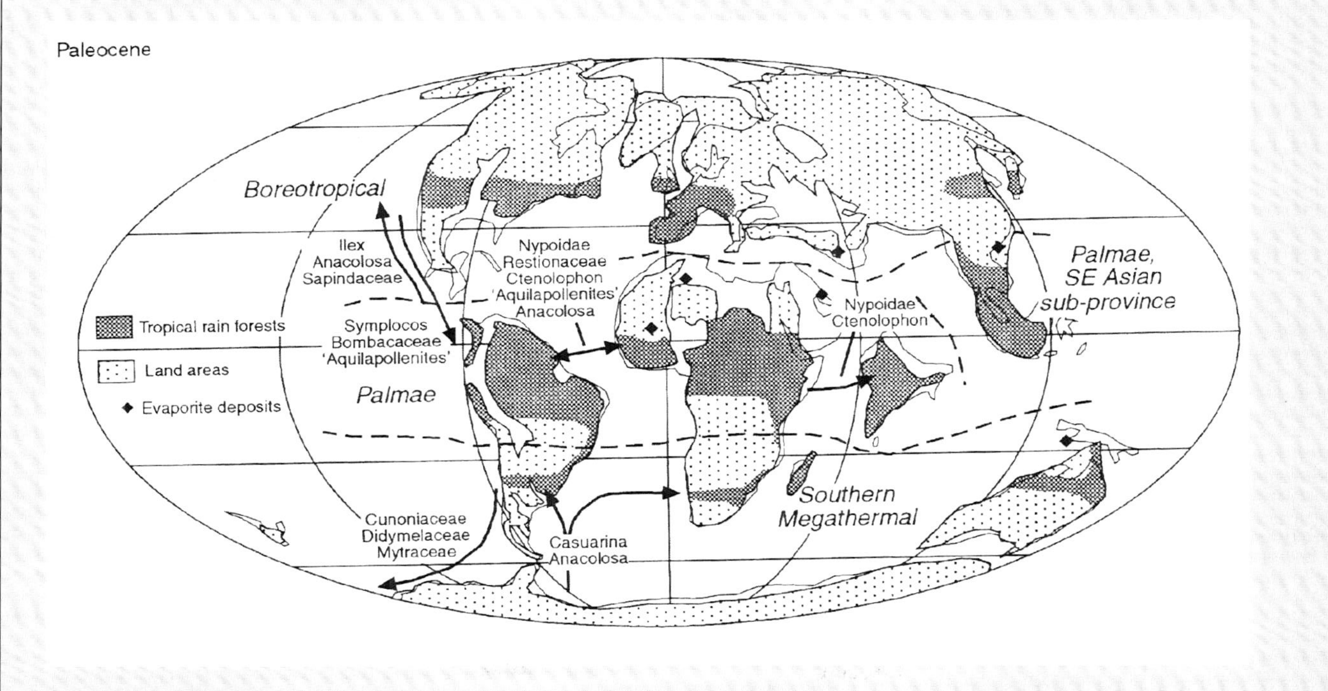

FIGURE 2.15 Reconstruction showing the large latitudinal range of tropical rainforests during the Paleocene greenhouse.
SOURCE: Morley (2000), courtesy of John Wiley & Sons, Inc.

be at least two orders of magnitude higher than the long-term average (Hassan et al., 2005), a rate potentially commensurate with the largest mass extinctions of the geological past (Sepkoski, 1996; Bambach, 2006). Modeling future biodiversity losses and their effects on the Earth's ecosystems and climate, however, is inherently difficult (Botkin et al., 2007), making it imperative to assess the outcome of equivalent "natural experiments" in the geological record (NRC, 1995; Myers and Knoll, 2001). The five major, and dozens of minor, mass extinctions of the past half-billion years (Sepkoski, 1996; Bambach, 2006) offer unique insights regarding ecosystem susceptibility and response to environmental stress, the potential for ecological collapse, and the mechanisms of ecosystem recovery (Benton and Twitchett, 2003; Bottjer et al., 2008). Furthermore, the integration of paleontologic, stratigraphic, and geochemical records for many intervals of the past half-billion years have revealed the variable character of past biotic turnovers and mass extinction events (e.g., Boxes 2.4, 2.6, 2.7, 2.8), which differ in regard not only to severity but also to duration, selectivity, and the nature of environmental stresses (e.g., the transition out of super-greenhouse conditions into Ordovician glaciation [Trotter et al., 2008]; the Early to Middle Triassic radiations [Payne et al., 2004]; the nannoplankton crisis and foraminiferal turnovers of the Cretaceous ocean anoxic events [Leckie et al., 2002]; Eocene-Oligocene faunal extinction and immigration [Kobashi et al., 2001; Ivany et al., 2004]). Most importantly, the geological record uniquely captures past climate-ecological interactions that are fully played out and thereby archive the impact, response, interaction, and recovery from past global warming and major climate transitions.

3

Climate Transitions, Tipping Points, and the Point of No Return

Because of the extended timescale—several centuries—necessary for climate to adjust to an increase in atmospheric CO_2, the current icehouse climate is out of equilibrium with long-term CO_2 forcing (Hansen et al., 2008). As the planet continues to warm, it may be approaching a critical climate threshold beyond which rapid (decadal-scale) and potentially catastrophic changes may occur that are not anticipated—because of complex feedback dynamics and existing computational limitations—by climate models that are tuned to modern conditions. This chapter focuses on the insights that can be gleaned from the deep-time geological archive of climate change concerning such thresholds, with particular focus on the major societal questions noted in Chapter 1: How soon, abrupt, and dramatic will climate change be, and how long will the new climate states persist?

Climate modeling efforts and the geological record provide plenty of evidence for climate system thresholds, or "tipping points" (Box 3.1), beyond which rapid changes can occur without any additional forcing (Hansen et al., 2008; Lenton et al., 2008). Components of the climate system that are particularly vulnerable to being forced by increasing atmospheric CO_2 across a threshold into a new state include the loss of Arctic summer sea ice, the stability of the Greenland and West Antarctic ice sheets, the vigor of the meridional overturning circulation in the North Atlantic and around Antarctica, the extent of Amazon and boreal forests, and the variability of the El Niño-Southern Oscillation (ENSO) (Lenton et al., 2008). The changes in state across such "tipping points" are typically accelerated relative to the apparent rate of forcing, are accompanied by large-scale

BOX 3.1
Tipping Points and the Point of No Return

There are sound theoretical reasons to think that tipping points across climatic thresholds exist (Gladwell, 2000; NRC, 2002). Examples of threshold behaviors include thermohaline circulation modifications, ice sheet instabilities, sea ice instabilities, soil-moisture feedbacks, and the onset of high-latitude convection and associated high-level cloud forcing. Hansen et al. (2008) introduced the term "tipping element" to describe subcontinental-scale subsystems of the Earth system that are susceptible to being forced into a new state by small perturbations. Tipping level—the magnitude of climate forcing beyond which, if sustained, abrupt climate change will eventually occur—is differentiated from "point of no return." If the tipping level is exceeded for only a brief period of time, the original state of the system can be restored. More persistent forcing can push the system to the "point of no return," where a reduction of the forcing below the tipping level is ineffective in halting the climate shift (Figure 3.1). This irreversibility of the system response is referred to as hysteresis (NRC, 2002).

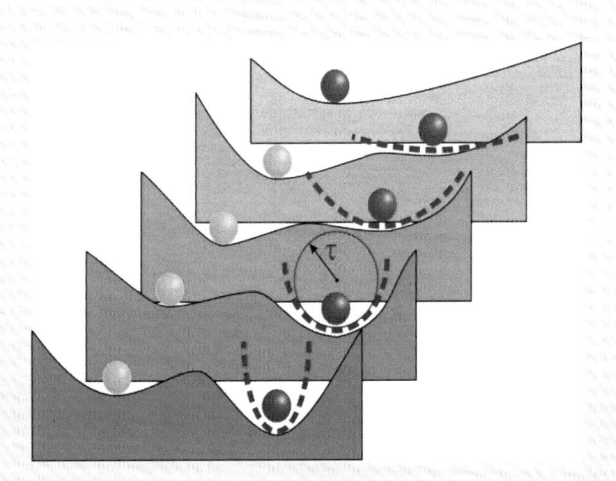

FIGURE 3.1 Equilibrium states of a "system" (valleys) in response to gradual anthropogenic CO_2 forcing (progressing from dark to light blue). The curvature of the valley is inversely proportional to the system's response time (τ) to small perturbations. A threshold is reached when the valley becomes shallower and finally vanishes causing the ball to abruptly roll to a new state (to the left).
SOURCE: Lenton et al. (2008), ©National Academy of Sciences, U.S.A.

impacts on ecological systems, and typically involve hysteresis (Lenton et al., 2008).

ICEHOUSE-GREENHOUSE TRANSITIONS

The following sections describe four periods of past climate change—icehouse-to-greenhouse or greenhouse-to-icehouse transitions—that were driven by slow (long-term) climate forcing across a critical threshold that led to abrupt and highly variable climate responses, as examples of what can be gleaned from the deep-time geological record of climate change and the scientific challenges that persist.

Initiation of the Cenozoic Icehouse

The early Cenozoic greenhouse Earth was plunged from a protracted state of warmth into its current glacial state 33.7 million years ago (Ma), at the Eocene-Oligocene boundary. The transition from a relatively deglaciated climate state to one in which the Antarctic ice sheet grew to between 40 and 160 percent of its modern size occurred within ~200,000 to 300,000 years (Coxall et al., 2005; Liu et al., 2009b). A long-term decrease in CO_2, commencing after the Early Eocene Climate Optimum at 52 Ma, has been proposed as the main cause of this cooling trend (Box 3.2) (Edmond and Huh, 2003; Kent and Muttoni, 2008). A CO_2 decrease through yet another apparent threshold (from as high as 415 ppmv [parts per million by volume] in the early Pliocene to ~280 ppmv; Pagani et al., 2010; Seki et al., 2010) most probably accounted for the initiation and growth of northern hemisphere ice sheets at around 3 Ma (DeConto et al., 2008; Lunt et al., 2008).

All of the elements of a tipping point climate transition are recorded by this greenhouse-to-icehouse turnover (Kump, 2009). As the climate system reorganized itself, it experienced an overshoot (the Oi-1 climate event) into a deep glacial, which was colder and with larger ice sheets than would be sustained during the less extreme conditions of the glaciated Oligocene (Zachos et al., 1996). The calcium carbonate compensation depth in the oceans deepened substantially in two 40-thousand-year (ky) long steps (separated by 200 ky) that occurred synchronously with the stepwise onset of major permanent ice sheets in Antarctica (Coxall et al., 2005). This instability in the climate system persisted for ~200 to 300 ky (Zachos et al., 2001b) and caused major changes in ocean and atmosphericic circulation with widespread effects on most marine and terrestrial ecosystems (Pearson et al., 2008). Such a characteristic response of a homeostatic feedback system implies an underlying dynamic that still remains to be fully understood but could result from changes in, and the interplay between,

BOX 3.2
Separating the Influence of Ice Volume and Temperature

The gradual cooling from the hothouse of the Early Eocene Climate Optimum (52 Ma) to the onset of Oligocene glaciation in Antarctica (~34 Ma) was first inferred from a long-term global trend of increasing benthic foraminiferal $\delta^{18}O$ values (Figure 3.2). The temperature of the deepest water in the oceans—an indication of global climate—was at least 10°C higher in the early Eocene than it is today. The cooling trend was disrupted several times by transient warming events in the Eocene and also by an abrupt shift toward heavier isotopic values (~1 to 1.5‰ [parts per thousand] increase in $\delta^{18}O$ in all records) at the Eocene-Oligocene boundary (referred to as the "Oi-1 overshoot"; Zachos et al., 2001b), with this transient cooling a result of some combination of rapid East Antarctic ice sheet growth and global cooling (Zachos et al., 2001a; Coxall et al., 2005). Marine carbon isotope compositions and $CaCO_3$ accumulation rates also exhibit the distinctive "overshoot," suggesting teleconnections between the southern hemisphere high latitudes and the tropical ocean (Coxall et al., 2005).

A number of additional proxies have been used to separate, or deconvolve, the effects of ice sheet growth from cooling—sequence stratigraphy to assess sea level change (e.g., Kominz and Pekar, 2001); marine geochemical proxies of temperature, including Mg/Ca ratios of foraminiferal calcite (e.g., Lear et al., 2000; Katz et al., 2008) and biomarkers (spores and pollen) in marine sediments (e.g., Liu et al., 2010); as well as terrestrial climate reconstructions based on oxygen isotopes in teeth and bones (e.g., Zanazzi and Kohn, 2008). The latest assessments indicate that the greenhouse-to-icehouse transition occurred in a series of steps with increasing influence of ice volume (Lear et al., 2008) and that cooling preceded ice sheet expansion, with maximum ice sheet size perhaps as much as 15 percent greater than today's Antarctic ice sheet (Pälike et al., 2006a; Liu et al., 2009b). A threshold was likely reached through a combination of orbitally driven changes in summer insolation and declining atmospheric CO_2 levels (DeConto and Pollard, 2003), although oceanic gateway opening and the thermal isolation of Antarctica may have played a role (Barker et al., 2007; Jovane et al., 2007).

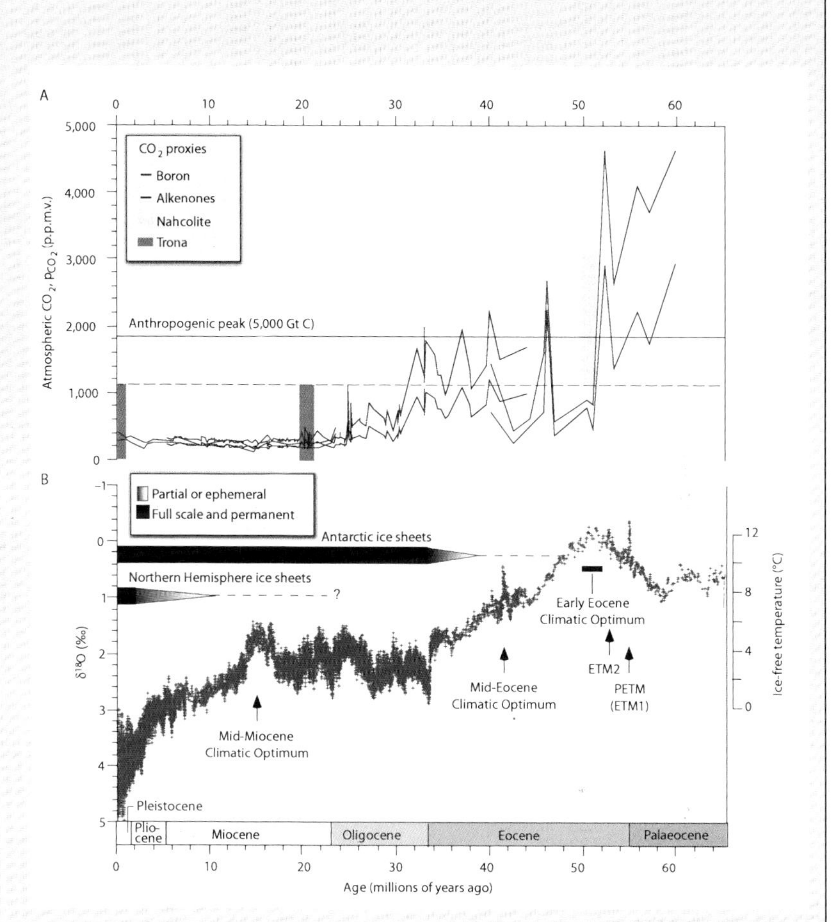

FIGURE 3.2 Relationship between atmospheric CO_2 (A) and climate (B) through the Cenozoic. The upper panel shows reconstructed pCO_2 from marine and lacustrine proxy records; the dashed line is maximum pCO_2 for the Neogene estimated by equilibrium calculations using lacustrine mineral phases (Lowenstein and Demicco, 2006). The climate curve in the lower panel is a composite of deep-sea benthic foraminiferal oxygen-isotope records, smoothed using a five-point running mean (Zachos et al., 2001a, 2008). The temperature scale on the right axis was calculated for an "ice-free ocean," and is thus applicable solely to the pre-Oligocene portion of record.
SOURCE: Zachos et al. (2008), reprinted by permission of Macmillan Publishers Ltd.

global silicate weathering rates, the global burial rates of marine $CaCO_3$ and siliceous plankton, atmospheric CO_2 levels, and ice sheet growth and ablation paced by changes in Earth's orbit (Coxall et al., 2005; Zachos and Kump, 2005; Pälike et al., 2006a).

The CO_2 threshold behavior exhibited by the Eocene-Oligocene onset of Antarctic glaciation and the Neogene initiation of the Greenland ice sheet suggests that multiple equilibrium states exist in the climate system. To the extent that the ice sheet climate system exhibits hysteresis, the CO_2 threshold identified for the cooling path may be substantially lower than that for the reverse warming path. Simulations of the modern climate system (DeConto and Pollard, 2003) and empirical proxy records (Pearson et al., 2009) have suggested a substantial delay (up to several millennia) in ice sheet response to increased atmospheric CO_2 due to hysteresis. Such studies indicate that polar ice sheet decay may require CO_2 levels well above those that existed during the initiation of Cenozoic glaciation (Pollard and DeConto, 2005). Recent evidence, however, indicates the potential for subdecadal response times of ice sheets and much more rapid melting (Das et al., 2008; van de Wal et al., 2008). The response of the Neogene polar ice sheets to the atmospheric CO_2 levels during the Middle Miocene climatic optimum (~500 ppmv at 16 Ma; Küerschner et al., 2008) and the early Pliocene (up to 415 ppmv at 4.5 Ma; Pagani et al., 2010)—values not too different from modern (2010) concentrations—warrants further exploration to resolve the uncertainties in ice sheet response times to global warming. If CO_2 forcing is sustained at levels through the point of no return, then rapid meltdown of glaciers can be anticipated in the future even if carbon emissions to the atmosphere ultimately decrease (Hansen et al., 2008).

If hysteresis is characteristic of ice sheet melting dynamics, then such a delay in ice sheet response to elevated CO_2 guarantees a future transition into a warm world that is abrupt, extreme, and with possibly irreversible catastrophic effects (Hansen et al., 2008; Kump, 2009). Presumably, the long-term processes that drove the climate system into the glacial state during the Cenozoic (enhanced silicate weathering and mountain building, reduced subduction of carbonates and volcanism, and thus low atmospheric CO_2 levels) will persist through the anthropogenic perturbation, so it is reasonable to anticipate that the climate—following the current transient warming—will cool over the subsequent few tens of millennia. Eventually, conditions for the reinitiation of the Antarctic and Greenland ice sheets will be achieved, but these may require atmospheric pCO_2 levels similar to preindustrial values and a favorable orbital state (Berger et al., 2003; Pollard and DeConto, 2005). Thus, the trip "forward to the past" may be quite prolonged, perhaps approaching the evolutionary timescales of species, including *Homo sapiens*.

Paleocene-Eocene Thermal Maximum (PETM)

One of the best-known examples of an ancient global warming event, with potential parallels to the near future, is the Paleocene-Eocene Thermal Maximum (PETM). This abrupt climate change occurred at ~56 Ma with repeated, rapid (millennial-scale), massive releases of "fossil" carbon and major disruption of the carbon cycle (Kennett and Stott, 1991, 1995; Dickens et al., 1995; Zachos et al., 2003). The oxygen isotopic compositions of planktonic and benthic foraminifera record rapid warming of ~5°C in tropical surface and deep oceans, and as much as 9°C warming at the poles, that persisted for ~170 ky (Sluijs et al., 2006; Zachos et al., 2006; Röhl et al., 2007) (Figure 3.3). Greenhouse gas-forced global warming was accompanied by extreme changes in hydroclimate and accelerated weathering (Bowen et al., 2004; Pagani et al., 2006; Schmitz and Pujalte, 2007), deep-ocean acidification (Zachos et al., 2005), and possible widespread oceanic hypoxia (Thomas, 2007; Zachos et al., 2008). Whereas regional climates in the mid- to high latitudes became wetter and were characterized by increased extreme precipitation events, other regions, such as the western interior of North America, became more arid (Schmitz and Pujalte, 2007). With this intense climate change came ecological disruption, including the immigration of modern mammalian orders (including primates) into North America, large-scale floral and faunal ecosystem migration (e.g., see Box 2.8), and widespread extinctions of benthic foraminifera in the deep ocean (Thomas and Shackleton, 1996; Bains et al., 1999; Wing et al., 2003, 2005). Carbon isotope records indicate that although the onset occurred within a few millennia, the recovery was much slower, taking well over 100 ky (Figure 3.3).

Dissociation (melting) of methane hydrates as their stability field crossed a threshold, triggered by a warming trend in the early Eocene (Dickens et al., 1995), is the most widely cited source of fossil carbon for the PETM. However, methane's isotopically light carbon requires that less carbon (~2000 petagram [Pg]) be added to account for the observed isotopic excursion than required by some models to account for the degree of inferred seafloor carbonate dissolution (Zachos et al., 2005; Panchuk et al., 2008). Some other suggested hypotheses to account for the abundant fossil carbon include sustained burning of accumulated Paleocene terrestrial organic peats and coals (Kurtz et al., 2003; Huber, 2008), although conclusive evidence in support of this hypothesis is lacking (Moore and Kurtz, 2008); increased terrestrial methane cycling (Pancost et al., 2007), although this may not generate a whole-system isotopic shift; desiccation and oxidation of organic matter in large epicontinental seaways (Higgins and Schrag, 2006), although the paleogeographic changes and their timing remain poorly resolved; and more speculatively, the impact of a volatile-rich comet (Cramer and Kent, 2005), although others have argued that the

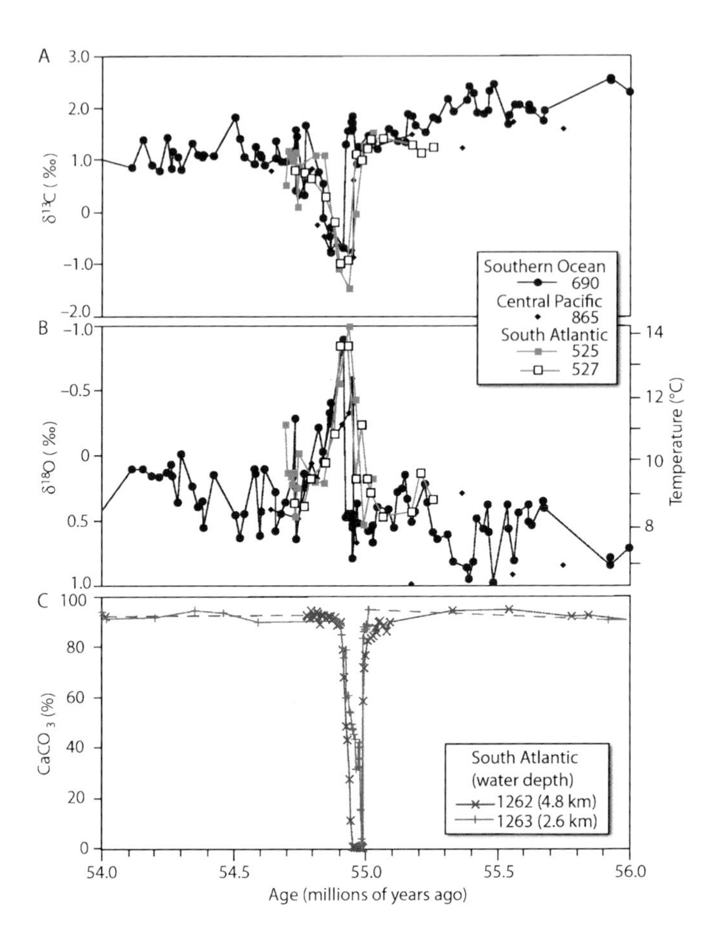

FIGURE 3.3 Marine stable isotope and seafloor sediment $CaCO_3$ records compiled using several ocean drilling sites for the PETM, a hyperthermal with some parallels to modern greenhouse gas-driven global change. (A) $\delta^{13}C$ time series developed from benthic foraminifera illustrating ~2.5 part per thousand (‰) excursion at ~55 Ma. (B) $\delta^{18}O$ time series and inferred temperatures record the prolonged period of ocean warming (~70-80 ky) and its large magnitude. There may have been several events of greenhouse gas release during the PETM that produced the large, abrupt changes in ocean temperatures. (C) Record of seafloor calcium carbonate content from the South Atlantic documents the significant reduction due to dissolution and deep-ocean acidification during the PETM. The apparent onset of $CaCO_3$ dissolution prior to the onset of the carbon isotope excursion reflects the extensive dissolution of uppermost Paleocene sediments by acidic waters during the PETM.
SOURCE: Zachos et al. (2008), reprinted by permission of Macmillan Publishers Ltd.

putative cometary particles were actually produced by bacteria (Kopp et al., 2007; Lippert and Zachos, 2007; Schumann et al., 2008). Clearly, there is no single fully satisfactory source to account for the carbon, and multiple carbon releases may have occurred in response to an initial warming. A likely trigger for the initial warming during the PETM is igneous intrusion into organic-rich sediments of the North Atlantic, which generated thermogenic methane and CO_2 (Svensen et al., 2004; Storey et al., 2007). Notably, the sudden release of carbon into the atmosphere-ocean system occurred at rates that vastly exceeded typical rates in Earth history, activating components of the climate system that can be triggered by accelerated warming. The PETM serves as an important base level showing the effect on the biosphere of a rapid rate of addition of fossil carbon to the atmosphere (~3,000 to 4,500 Pg—on the order of that anticipated if we burn through all fossil fuels)—yet dwarfed by the present rate of ~1 percent per year CO_2 increase in Earth's atmosphere (Zeebe et al., 2009).

Transient warming episodes, such as the PETM, were a recurring phenomenon of the early Eocene warm world. Many of the short-lived hyperthermals were associated with abrupt and extreme climate change, an accelerated hydrological cycle, and ocean acidification (Nicolo et al., 2007; Stap et al., 2009) (Box 3.3). Short-term positive feedbacks active during the hyperthermals magnified the climatic effects of the initial carbon influx. Climate amelioration with each transient warming event would have been substantially delayed as the rates of short-term feedbacks far outpaced the negative feedbacks (e.g., weathering) capable of restoring the global carbon cycle to a steady state (Zachos et al., 2008; Zeebe et al., 2009).

The PETM, and other hyperthermals of the early Cenozoic, occurred when the Earth was virtually ice-free. This is certainly significantly different from modern and near-future conditions, which are expected to maintain unipolar glaciation at a minimum. The ice sheets of the Neoproterozoic Snowball Earth and the Late Paleozoic Ice Age were far more extensive than those of the Cenozoic icehouse, recording repeated major glacial-interglacial transitions and including terminal epic deglaciation. Despite substantially different land mass-height distributions, ocean circulation patterns, and marine and terrestrial ecosystems from those of today, the geological record of these deglaciations—specifically the repeated major transitions between glaciations and glacial minima including their terminal epic deglaciations—provide the only "icehouse" perspective of the response of the climate system and ecosystems to perturbation beyond the range archived in the more recent glacial records.

The Late Paleozoic Deglaciation

Much of the scientific understanding of feedbacks and thresholds in the current glacial climate system, and their influence on the biosphere,

BOX 3.3
Deep-Time Insights into Ocean Acidification

The early Cenozoic hyperthermals, and in particular the PETM, provide a natural laboratory to study climate sensitivity to pCO_2, the interplay of short- and long-term feedbacks in the climate system, and ocean acidification under magnitudes of atmospheric pCO_2 increases that are comparable to present and projected future increases. Clear evidence of deep-ocean acidification exists for the PETM (Zachos et al., 2005; Zeebe and Zachos, 2007), with corrosive waters completely dissolving calcium carbonate on the Atlantic seafloor at water depths below 2.5 km; today, this calcium carbonate compensation depth (CCD) occurs below 4 km in most ocean basins.

Whether surface waters became undersaturated is less clear since carbonate producers such as planktonic foraminifera and coccolithophorids persisted during the event. However, whereas shallow water reefs composed of corals, calcareous red and green algae, and larger benthic foraminifera were abundant prior to the PETM, metazoan reefs nearly vanished between 56 million and 55 million years ago (Scheibner and Speijer, 2008; Kiessling and Simpson, 2010). Widespread coral reefs did not reappear until the middle Eocene, at ~49 Ma. It appears that a combination of persistent warming from the late Paleocene to early Eocene, punctuated by deep-ocean acidification at the PETM, defined a threshold for coral-algal reefs that led to rapid loss and only gradual recovery. Notably, the lack of evidence for surface water acidification probably indicates that the rate of carbon addition was slower—perhaps by an order of magnitude—than projected fossil fuel emission rates under the least optimistic scenarios for the future (e.g., the A1 family of scenarios considered by IPCC [2007]) which, in box models, generates surface *and* deep-water acidification (Zeebe et al., 2008, 2009).

Over the past two centuries, the ocean has absorbed 40 percent of anthropogenic CO_2 emissions (Zeebe et al., 2008). If fossil fuel emissions continue unabated and minimal development is put into carbon sequestration technologies, by the time humans burn through estimated fossil fuel reserves (at ~A.D. 2300 to A.D. 2400), ~5,000 gigatonnes of carbon will have been released to the atmosphere (Zachos et al., 2008). Because the rate of anthropogenic carbon input to the atmosphere greatly exceeds the mixing time of the oceans (1,000-1,500 years), CO_2 will build up in the atmosphere (perhaps to ~2,000 ppmv) and the surface ocean (Kump, 2002; Zachos et al., 2008). What could be in store for this millennium? As the ocean continues to absorb CO_2, carbonate ion (CO_3^{2-}) concentration will fall leading to decreases in surface water pH and saturation states, a condition that is already apparent and will continue over the next century (Figure 3.4). Acidic surface waters are expected to massively affect ocean ecosystems, including the widespread loss of coral reefs. With time, acidic

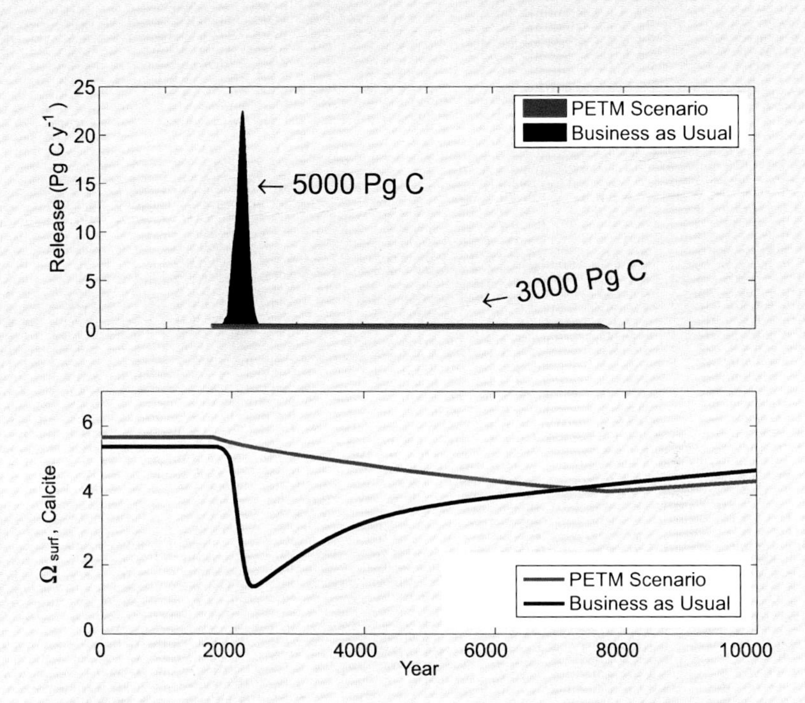

FIGURE 3.4 Initial carbon pulse for the PETM (red curves), estimated to be 3,000 Pg carbon using published carbon isotope and observed deep-sea carbonate dissolution records, and a carbon cycle model (LOSCAR; Zeebe et al., 2008, 2009). The magnitude of the input carbon mass was inferred from carbonate dissolution records, with the $\delta^{13}C$ of the carbon pulse (≤ –50‰) constrained by requiring the model outcome to match observed deep-sea $\delta^{13}C$ records. The model assumes a large initial input of carbon over 5 ky, followed by further smaller pulses and a low continuous carbon release (an additional 1,500 Pg) throughout the PETM main event. Changes in calcite saturation in the surface ocean (lower diagram) are estimated for the PETM (red curve) and for the future (black curve), based on the inferred magnitude of the carbon pulse to atmosphere. SOURCE: Courtesy of R.E. Zeebe, personal communication (2010).

continued

BOX 3.3 Continued

water will penetrate to the deep ocean where it will dissolve carbonate sediments and begin to be neutralized. In response, the saturation horizon of the deep ocean will shoal on a decadally observable timescale. Calcification rates of corals will slow noticeably and may become negligible in the next 100-150 years. At first, the rise in the saturation horizon will be slow, but as the area of seafloor above the saturation horizon declines (following the seafloor hypsometric curve), the shoaling rate will accelerate, bringing it to as shallow as the depth of the shelf-slope break (~130 m) in the next several centuries. At this point, barrier reefs, having long since lost their reef-building biota, will erode through dissolution and disintegrate. This history of carbonate dissolution will result in a carbonate-poor layer in the deep ocean, much like sediments associated with past hyperthermals such as the early Cenozoic PETM.

has been elucidated by studies of climate transitions of the past few million years—in particular the moderate-scale glacial-interglacial fluctuations of the Pleistocene. The demise of the Late Paleozoic Ice Age (LPIA; between 290 and 260 Ma) provides an opportunity to evaluate climate stability and climate-biota interactions during a major climate transition coupled to changing CO_2 contents. For example, climate models indicate that climate-driven biome changes at high latitudes may have factored strongly in controlling LPIA glacial-interglacial changes (e.g., Horton et al., 2010). For the final stages of this protracted ice age, covariance between shifts in pCO_2 and continental and marine surface temperatures inferred from isotopic proxies of soil-formed minerals and marine fossil brachiopods, and ice sheet extent reconstructed from southern Gondwanan glacigenic deposits, indicates a strong linkage of pCO_2-climate-ice-mass dynamics that is consistent with greenhouse gas forcing (Montañez et al., 2007). A pattern of progressively more extensive and long-lived ice sheets through the Late Carboniferous (340 to 310 Ma; Fielding et al., 2008) was reversed in the Early Permian—under rising atmospheric CO_2 levels—as climate ameliorated and conditions shifted toward a protracted greenhouse climate state (Montañez et al., 2007). The trend of gradually increasing surface temperatures and increasing atmospheric pCO_2 is punctuated by larger but shorter-term fluctuations associated with each discrete glaciation. Surface temperatures and CO_2 levels never returned to the earliest Permian minima associated with the apex of Gondwanan continental ice sheets. Intermittent warmings—characterized by CO_2 levels above the simulated threshold for glaciation (Horton et al., 2007; Horton and Poulsen, 2009)—heralded the more permanent change to an ice-free world to come. This pattern of episodic changes in atmospheric CO_2 and surface temperatures in step with transient glaciations, superimposed on a longer-term warming trend during the demise of the Late Paleozoic Ice Age, shares characteristics in common with the Eocene-Oligocene greenhouse-to-icehouse transition (Coxall et al., 2005; Liu et al., 2009a,b). The similarity in behavior of these very different transitions suggests that turnovers in climate states are most probably characterized by large-scale episodic change. Future study of the deep-time record of the Earth's last epic deglaciation should shed light on how the cryosphere, hydrosphere, chemosphere, and biosphere responded to such episodic change under rising CO_2 levels.

Deglaciation During the Neoproterozoic

A phase of rapid global warming is recorded in the late Neoproterozoic (~635 Ma), abruptly terminating what was probably the longest-lived (~135 Ma; Macdonald et al., 2010) and coldest icehouse period of Earth history, where at times ice sheets extended to sea level in equatorial latitudes—a climate state popularly referred to as the "Snowball Earth" (Hoffman et al., 1998). Carbon isotope trends provide evidence for substantial, but poorly understood, disruption of the carbon cycle during the ice age itself, including the possibility that the high albedo of a global-scale ice sheet dominated climate (Hoffman et al., 1998). The terminal deglaciation in the Neoproterozoic offers an intriguing deep-time archive of how major changes in long-term processes that regulate climate, such as silicate weathering and carbon burial and productivity, have been triggered when a threshold in the climate system has been reached through CO_2 forcing. Abrupt and rapid increase in CO_2 at the end of the Neoproterozoic glaciations is recorded by the presence of thin calcium carbonate deposits, interpreted to have been deposited on a millennial timescale, immediately overlying Neoproterozoic glacial sediments across the globe (Kennedy et al., 1998; Hoffman and Schrag, 2002). These carbonate deposits are a physical record of rapid release of CO_2 to the atmosphere calculated at a rate of ~1 percent CO_2 increase per year (Kennedy et al., 2001), similar to the current rate of CO_2 increase of 0.8 to 1 percent per year (IPCC, 2007). While the analogy is imperfect because of the very different biosphere and continental configuration in the Neoproterozoic, deglaciation under this strong greenhouse gas forcing imparted a record unique from that of subsequent deglaciations. Most notably, the abrupt transition to greenhouse conditions associated with this complete deglaciation appears to have involved a dominating rapid-warming feedback (Fairchild and Kennedy, 2007) that involved the massive release of CO_2 (Hoffman and Schrag, 2002) and/or the destabilization of methane clathrates and the release of methane gas from permafrost and marine reservoirs (Kennedy et al., 2001, 2009). It appears that the dominance by warming feedbacks during the termination of this ice age precluded the damping effects of other feedbacks that governed climate oscillations during Phanerozoic ice ages. Consequently, the late Neoproterozoic deglaciation provides an excellent example of the long-term feedbacks that can be triggered—and likely accelerated—when a threshold in a strongly forced climate system is reached.

HOW LONG WILL THE GREENHOUSE LAST?

The potential for climatic consequences with severe impact on humans resulting from the buildup of fossil fuel CO_2 has inevitably resulted in questions not only of "How bad will it get and how fast?" but also "How long will it last?" The answers to these questions depend heavily on the global warming potential of the greenhouse gas release, which accounts not only for its immediate impact on the planetary radiative energy balance but also on the longevity of the greenhouse gas in the atmosphere (Archer et al., 2009).

Economic forecasts suggest that conventional fossil fuel resources will largely be used up in the next 200-400 years, leading to atmospheric CO_2 levels that could reach ~2,000 ppmv by A.D. 2300 to 2400 (Marland et al., 2002; Caldeira and Wickett, 2003). However, models of the global carbon cycle and the geologic record both show that CO_2 produced from fossil fuels and other reservoirs will continue to impact global climate and atmospheric chemistry for tens to hundreds of thousands of years. Although CO_2 produced by fossil fuel burning is taken out of the atmosphere within decades of its production, the oceans, soils, and vegetation continue to exchange greenhouse gases back into the atmosphere for far longer. Greenhouse gases continue to affect climate and ocean acidity until they are buried as organic matter or converted to mineral forms of inorganic carbon through rock weathering (Box 3.4).

Simple box models (Figure 3.6) have been used to make long-term projections of future climate to capture the "recovery" from the fossil fuel-induced greenhouse state (Walker and Kasting, 1992; Archer, 2005). Although box model calculations should not be considered definitive, they do suggest that the fossil fuel perturbation may interfere with the natural glacial-interglacial oscillation driven by predictable changes in Earth's orbit (Berger et al., 2003), perhaps forestalling the onset of the next northern hemisphere "ice age" by tens of thousands of years. A more convincing exposition of the central question of "how long" requires more comprehensive models. Scientific confidence in those models will be high

BOX 3.4
CO_2 Sweepers and Sinks in the Earth System

The carbon fluxes in and out of the surface and sedimentary reservoirs over geological timescales are finely balanced, providing a planetary thermostat that regulates Earth's surface temperature. Initially, newly released CO_2 (e.g., from the combustion of hydrocarbons) interacts and equilibrates with Earth's surface reservoirs of carbon on human timescales (decades to centuries). However, natural "sinks" for anthropogenic CO_2 exist only on much longer timescales, and it is therefore possible to perturb climate for tens to hundreds of thousands of years (Figure 3.5). Transient (annual to century-scale) uptake by the terrestrial biosphere (including soils) is easily saturated within decades of the CO_2 increase, and therefore this component can switch from a sink to a source of atmospheric CO_2 (Friedlingstein et al., 2006). Most (60 to 80 percent) CO_2 is ultimately absorbed by the surface ocean, because of its efficiency as a sweeper of atmospheric CO_2, and is neutralized by reactions with calcium carbonate in the deep sea at timescales of oceanic mixing (1,000 to 1,500 years). The ocean's ability to sequester CO_2 decreases as it is acidified and the oceanic carbon buffer is depleted. The remaining CO_2 in the atmosphere is sufficient to impact climate for thousands of years longer while awaiting sweeping by the "ultimate" CO_2 sink of the rock weathering cycle at timescales of tens to hundreds of thousands of years (Zeebe and Caldeira, 2008; Archer et al., 2009). Lessons from past hyperthermals suggest that the removal of greenhouse gases by weathering may be intensified in a warmer world but will still take more than 100,000 years to return to background values for an event the size of the Paleocene-Eocene Thermal Maximum.

In the context of the timescales of interaction with these carbon sinks, the *mean* lifetime of fossil fuel CO_2 in the atmosphere is calculated to be 12,000 to 14,000 years (Archer et al., 1997, 2009), which is in marked contrast to the two to three orders of magnitude shorter lifetimes commonly cited by other studies (e.g., IPCC, 1995, 2001). In addition, the equilibration timescale for a pulse of CO_2 emission to the atmosphere, such as the current release by fossil fuel burning, scales up with the magnitude of the CO_2 release. "The result has been an erroneous conclusion, throughout much of the popular treatment of the issue of climate change, that global warming will be a century-timescale phenomenon" (Archer et al., 2009, p. 121).

continued

BOX 3.4 Continued

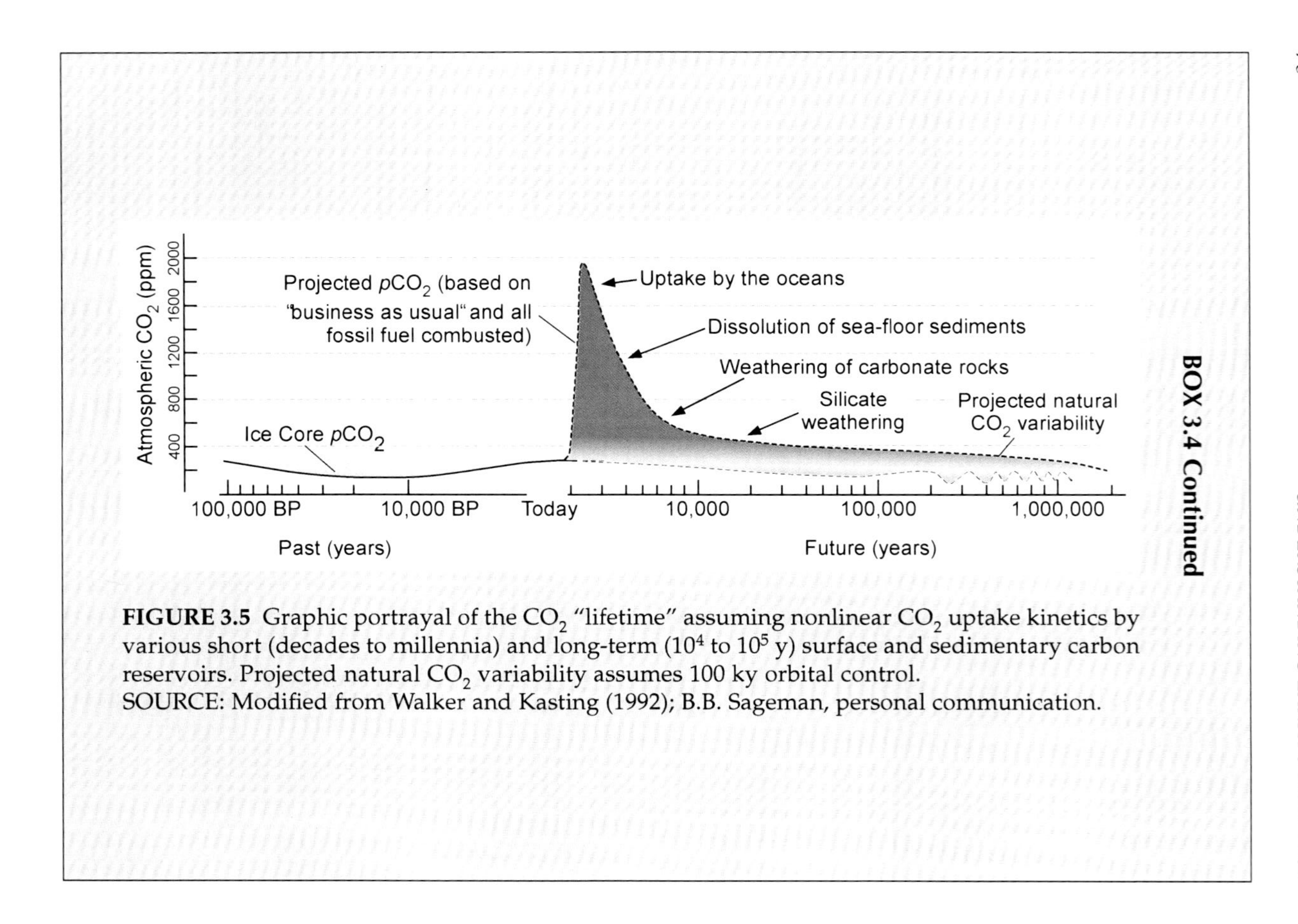

FIGURE 3.5 Graphic portrayal of the CO_2 "lifetime" assuming nonlinear CO_2 uptake kinetics by various short (decades to millennia) and long-term (10^4 to 10^5 y) surface and sedimentary carbon reservoirs. Projected natural CO_2 variability assumes 100 ky orbital control.
SOURCE: Modified from Walker and Kasting (1992); B.B. Sageman, personal communication.

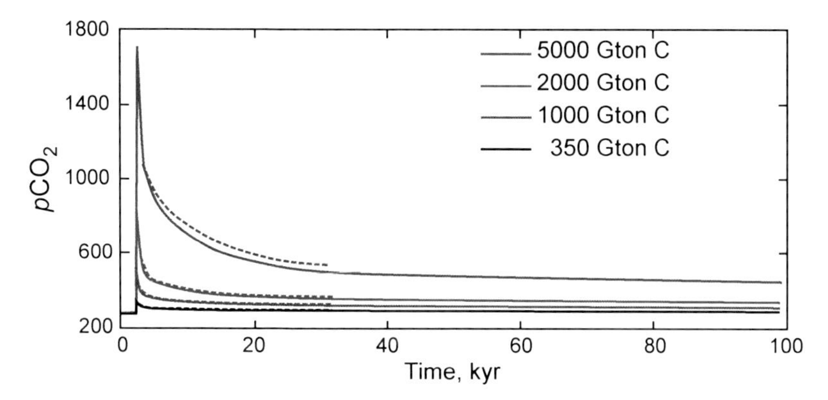

FIGURE 3.6 Schematic showing the predicted long-term response over 100,000 years of atmospheric CO_2, including ocean temperature feedbacks, to a range of possible fossil fuel emissions totals. The 100,000-year simulations include silicate weathering (solid lines) and the 35,000-year simulations include seafloor $CaCO_3$ dissolution (dashed lines). These models highlight how the timescale of carbon uptake becomes extended as the event unfolds. Fast processes such as ocean uptake and biomass growth, with high transfer rates but limited capacity, lose their potency, while slower processes, such as seafloor carbonate dissolution and rock weathering, come to dominate.
SOURCE: Modified from Archer (2005).

only if they can be evaluated against observation. The historical record, and even the expanse of the Quaternary climate record, contains nothing comparable.

Observations and modeling of the past carbon cycle perturbations provide a basis for projecting future conditions under the full range of fossil fuel burning scenarios, including the most pessimistic "business-as-usual" eventuality. Along this trajectory, atmospheric CO_2 levels will rise as long as fossil fuel burning continues (with ultimate input of ~5,000 GtC), rising to levels perhaps as high as 1,600-2,000 parts per million (i.e., five to seven times the preindustrial level) (Figure 3.6). The geological record of past hyperthermal events, including the PETM, suggests that severe global warming under such magnitudes of carbon emissions will persist for 20,000 to 40,000 years. Carbon cycle models indicate that even after 100,000 years, the anthropogenic perturbation to the carbon cycle will still be important, especially if the total amount of carbon emitted is large. Consequently, Milankovitch forcings that have so dominated the pacing and extent of climate variations, and especially ice sheets, over the last 2 million years will—as they were prior to the onset of the current glacial state in the Oligocene—serve as only a minor modulator of high-latitude climate variability because climate change will be muted under such elevated atmospheric pCO_2 levels. The Greenland ice cap could disappear in the first few coming millennia, and if CO_2 levels rise more than two to four times present levels, the West Antarctic ice cap could collapse (Naish et al., 2009), although this conclusion is highly sensitive to orbital configuration and model parameterizations (Pollard and DeConto, 2005). By any measure, exploitation of much of the fossil fuel reservoir over only 300 years will clearly leave a far longer lasting legacy.

4

Deciphering Past Climates—Reconciling Models and Observations

Forecasts of climate change for the next century are based on general circulation models (GCMs) that have been developed and tuned primarily using twentieth century records, but also with some input based on the understanding of the climate system from the more recent geological past (e.g., the Last Glacial Maximum and the Holocene). In part, this reflects the high levels of radiometric calibration and temporal resolution (subannual to submillennial) offered by near-time paleoclimate archives, which are capable of identifying the typically nonlinear components of the climate system—characterized by rapid response times—that are relevant to human society. A critical prerequisite for accurate forecasts of future regional and global climate changes based on GCMs, however, is the requirement that these models use parameters that are relevant to the future we seek to better understand. In this context, the recent climate archive captures only a small part of the known range of climate phenomena, since it has been derived from a time dominated by ice dynamics at both poles. Furthermore, the magnitude of forcing that the planet is now experiencing exceeds any that has occurred during the past ~30 million years. As GCMs are transformed into Earth System Models for the Intergovernmental Panel on Climate Change (IPCC) Fourth Assessment Report, they will encompass vastly more system physics, and deep-time climate studies will offer modelers the only real-world scenarios for testing the full climate response to the large increases in greenhouse gas levels that are projected.

As the climate system departs from the conditions captured by these well-studied near-time climate analogues, it is necessary for the scientific community's efforts to expand to capture the full range of variability and climate-forcing feedbacks of the global climate system, in particular for the past "extreme climate events" and warmer Earth intervals that may serve as analogues for future climate. Full testing of climate models for these time periods will require evaluation of feedback processes within models, enhanced spatial resolution, and longer simulations to better characterize climate model variability. All of these requirements, especially those for resolution and variability, will require significant computational resources.

For deep-time climate systems, the representation of paleogeographic boundary conditions can be a much greater source of uncertainty than it is for simulations based on modern geography. Furthermore, discrepancies between model outputs and paleoclimate observations may indicate the existence of additional processes, feedbacks, and/or sensitivities that are not present in the model or expressed in the modern climate system. For example, the exceptionally warm high latitudes during all past warm periods—whether transient or long term—cannot be reproduced by models without invoking unreasonable CO_2 levels, revealing the inability of current models to fully capture the processes and feedbacks governing heat transport and retention or the processes that might generate heat in the polar regions under elevated atmospheric greenhouse gas levels (Covey and Barron, 1988; Rind and Chandler, 1991; Covey, 1991; Sloan and Pollard, 1998; Bice et al., 2006; Huber, 2008; Kump and Pollard, 2008; Spicer et al., 2008; Zachos et al., 2008). Thus, model development, which is based on improving specific processes and climate feedbacks and, in turn, evaluating the impact of these improvements on model simulations, depends on the availability of spatially resolved, robust, deep-time paleoclimate reconstructions of appropriate geochronological resolution and constraint. In addition, the utility of paleoclimate proxies for climate reconstruction and data-model comparisons relies on the proxies being sufficiently well preserved and the existence of an adequate understanding of the underlying processes, sensitivities, and uncertainties captured by these proxies.

Recent paleoclimate studies of deep-time successions have documented the potential of the older part of the geological record to reveal long-duration archives of forcings, responses, and long-term (centuries to tens of millennia) feedbacks that are of magnitudes and/or durations not captured by Pleistocene and Holocene paleoclimate records. Constraining the nature (e.g., rates, phasing between proxies) and origin (forcings) of climatic shifts, particularly rapid and/or transient events across climate thresholds from the deep-time record, will be greatly enhanced where orbital-scale cycles can be identified and resolved in the rock record (Box 4.1, Figures 4.1 and 4.2). Indeed, millennial to seasonal signals—at times calibrated to the orbital timescale—have been extracted from the sedimentary record spanning hundreds of millions of years (e.g., Feldman

BOX 4.1
Determining Time in the Geological Record

One of the biggest challenges in using the deep-time record for understanding Earth systems is determining the rates of processes and dating when specific events occurred. Determining rates requires very precise time control, particularly if the processes being studied occur at an ecological timescale (1 year to several centuries). One method for such precise time control is by using annually layered sediments in ancient anoxic or hypersaline basins, as long as a few age control points are present. Examples include laminated sediments from the Pleistocene Santa Barbara Basin (Figure 4.1a), Eocene sediments of the North Sea (Figure 4.1b) (Schiøler et al., 2007), black shale sequences in the Cretaceous North Atlantic (Figure 4.1c), and the Permian Castile Formation of West Texas (Anderson, 1982). Quite highly resolved relative timescales can also be achieved using cyclic sequences, with resolutions of a few centuries to several tens of millennia, based on the identification of distinct orbital periods (Figure 4.1d; Figure 4.2). Both annual layers and orbitally tuned records can reveal ecological dynamics, snapshots of natural variability at different parts of Earth history, and the duration of threshold shifts in ecosystems.

Orbital cycles can sometimes be tied to astronomically tuned timescales during the past 40 million years to provide excellent absolute timescales. One example is the tuning of orbital cycles in glacial events during the Oligocene using combined geochemical and sediment property cycles tied to an astronomically calibrated timescale (Pälike et al., 2006a,b; see Figure 4.2). Orbital cycles have been recognized far back in the Phanerozoic sedimentary record and, together with high-resolution U-Pb dating, offer the potential to reconstruct Earth system dynamics in great detail (Erwin 2006; Davydov et al., 2010).

et al., 1993; Olsen and Kent, 1996; Eriksson and Simpson, 2000; Loope et al., 2001, 2004; Ivany et al., 2004; Wagner et al., 2004; Elrick and Hinnov, 2007; Jahren and Sternberg, 2008; Kennedy et al., 2009). The ability to precisely and accurately quantify geological time has improved dramatically with recent advances in radiometric dating and interlaboratory cross-calibration (e.g., the EARTHTIME initiative) permitting unprecedented temporal resolution (e.g., ID-TIMS [isotope dilution-thermal ionization mass spectrometry] uranium-lead [U-Pb] ages with analytical uncertainties of ≤0.01 percent; Ramezani et al., 2007). Some recent radiometric calibrations of the sedimentary record integrate astrochronology, providing Milankovitch-scale resolution through long intervals of time

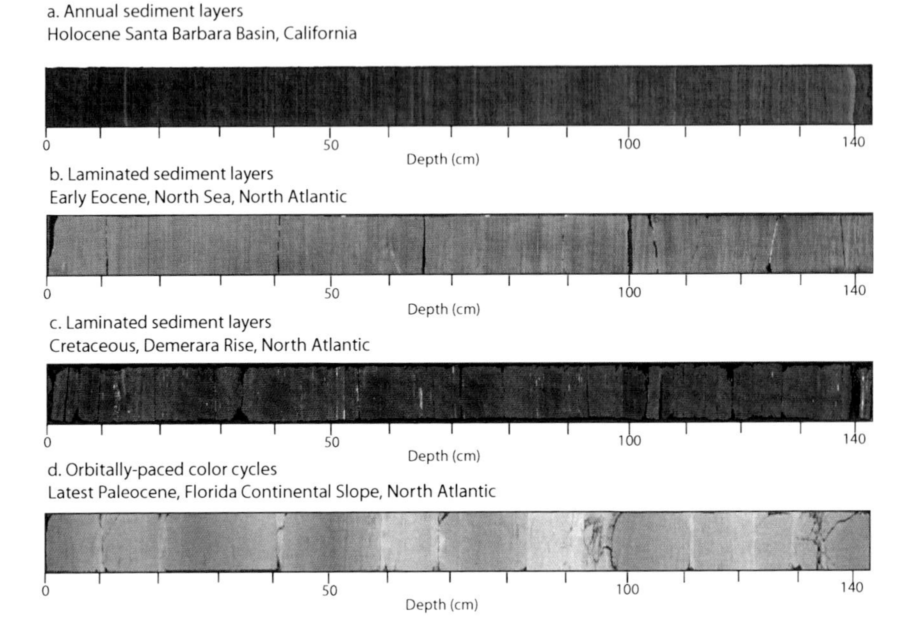

FIGURE 4.1 Sediment cores can be detailed recorders of time: (a) Core sample (Core MV1012-TC-3) from the Holocene of Santa Barbara Basin, California, displaying well-developed millimeter-scale laminations related to annual cycles in biological productivity and sediment runoff. (b) Eocene laminated sediments from the North Sea (Nini-3 well), potentially offering very high resolution records of climate and ecosystem variability. (c) Similar laminations in a drill core through Cretaceous black shales from the tropical North Atlantic (from ODP Site 1259). (d) Orbital cycles in sediment color paced by the precession cycle (~21 kyr) from a drill core in the Paleocene of the western North Atlantic (from ODP Site 1051; Norris and Röhl, 1999).
SOURCE: ODP core images courtesy Integrated Ocean Drilling Program Science Services.

(e.g., Kuiper et al., 2008; Davydov et al., 2010). Furthermore, integration of orbital-stratigraphic approaches with bio-, magneto-, cyclo-, and/or chemostratigraphy has successfully placed high-resolution temporal constraints on past events (e.g., Olsen and Kent, 1996; Sageman et al., 2006; Westerhold et al., 2008; Adams et al., 2010; Galeotti et al., 2010). Several epochs and stages of the Phanerozoic have been fully orbitally tuned, presenting the possibility of geochronological resolution at 10^4- to 10^5-year scales (e.g., Hinnov and Ogg, 2007). Many of these records, however, await radiometric calibration to the absolute timescale.

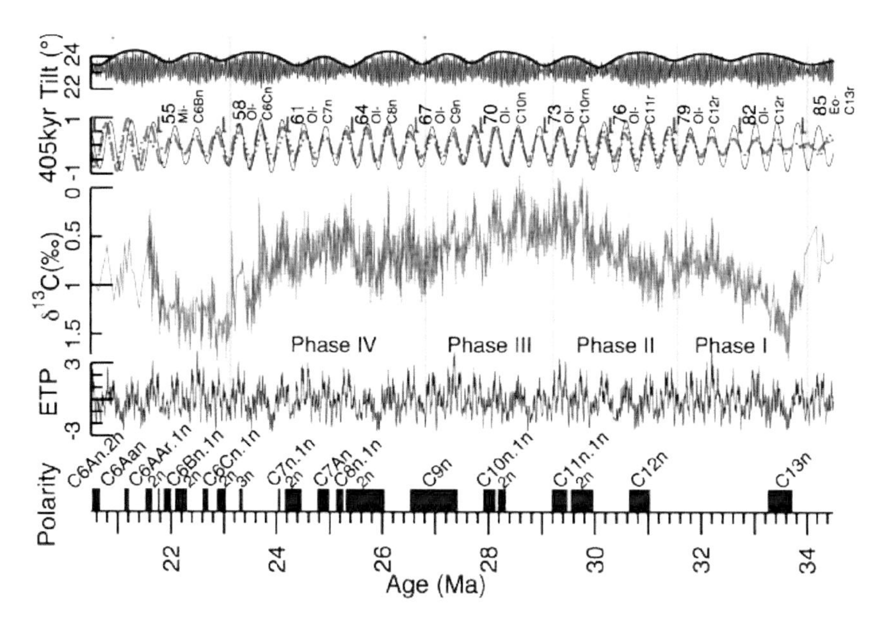

FIGURE 4.2 Astronomically tuned climate record from the Oligocene of the Central Pacific. Top two panels show the astronomical calculation of the 41,000-year cycle in the tilt of Earth's spin axis (known as obliquity), and the 405,000-year cycle in eccentricity (thin black line). In the middle panel is the carbon isotope record of the deep Pacific which displays a well developed ~400,000-year cycle that closely matches the calculated astronomical cycle (dashed green line). ETP is the calculated cycle expected if the sediment record integrated the combined eccentricity, tilt and precession cycles. The geomagnetic polarity timescale is shown on the bottom, here calibrated to the astronomical cycles.
SOURCE: Modified from Pälike et al. (2006b).

The potential of the deep-time paleoclimate record to provide unique insight into scientific understanding of the climate system's response to greenhouse gas forcing, however, is underdeveloped. To maximize this potential, the community is presented with three primary challenges:

- To determine precise chronologies for existing and new geological archives of paleoclimate interest—where feasible at the temporal resolutions that are possible in the Pleistocene and Holocene—through continued improvements in the precision and accuracy of geochronological techniques applicable to the sedimentary record (Ar-Ar [argon-argon] and U-Pb), and the development of novel radiometric approaches such as U-Pb dating of carbonates (Rasbury and Cole, 2009) and rhenium-osmium

(Re-Os) dating of black shales (Ravizza and Turekian, 1989; Selby and Creaser, 2005; McArthur et al., 2008a).

- To develop within such geochronological and/or orbitally tuned frameworks, marine and terrestrial—ultimately linked—time series of high temporal resolution and spatial density and distribution. For terrestrial records that are notoriously geographically fragmented and poorly resolved, this will rely in large part on systematically acquired, targeted continental drilling of continuous and highly resolved records.
- To obtain proxies of a broad range of surface and atmospheric conditions (paleotemperatures, pCO_2, pO_2, contents of other greenhouse gases, paleoprecipitation, seasonality, relative humidity)—of higher precision and accuracy than currently available—through some combination of proxy refinement, proxy development, and multiproxy studies.

CLIMATE MODEL CAPABILITIES AND LIMITATIONS

Current paleoclimate model capabilities (Box 4.2) include the application of fully coupled three-dimensional models to past climates. These are the same models that are used to study the present climate state and future changes to Earth's climate, and they are the models that provide the modeling foundation for the IPCC periodic assessments. For many years, these models included only the physical aspects of the atmosphere, dynamic ocean, land, and sea ice components of the climate system. More recently, however, these models have begun to include coupling to a dynamic ice sheet model and prognostic components for biogeochemistry, atmospheric chemistry, dynamic vegetation, and ecology. Many models can even provide calculations of the isotopic composition of precipitation, making direct comparisons with $\delta^{18}O$ marine and terrestrial proxies possible (Roche et al., 2006; Poulsen et al., 2007a, 2010; Zhou et al., 2008). Thus, global climate models have evolved from physical climate system models to more comprehensive Earth system models that permit more realistic coupling between the physical climate system and the biosphere (e.g., Slingo et al., 2009; Cadule et al., 2010).

In terms of the mean state, climate system models are able to realistically capture many characteristics of the current climate, such as observed equator-to-pole thermal gradients, large-scale spatial distribution of precipitation patterns, and various aspects of regional climate variability (e.g., El Niño-Southern Oscillation, Pacific Decadal Oscillation). Although the more comprehensive Earth system models offer many advantages, many aspects of regional-scale climates are still not captured accurately (although see Sewall and Sloan [2006], Thrasher and Sloan [2009], for exceptions). Simulating regional-scale phenomena requires the existence of high-resolution paleoclimate boundary data, which may not exist for

BOX 4.2
Climate Models

Climate models are numerical representations of the climate system that provide a means to study the processes that determine Earth's climate state. Over the past 50 years, climate models have evolved to include a hierarchy of approaches to representing the climate system (Figure 4.3).

Geochemical box models are used to study the temporal evolution of quantities such as atmospheric CO_2 and oxygen, ocean stable isotopes, and other geochemical variables. These models are based on theoretical expressions of the sources and sinks of a range of geochemical properties, providing global mean information on timescales of tens of thousands to millions of years.

Earth system models of intermediate complexity (EMICs) extend the box model concept to include spatial resolution and are useful tools to study Earth processes for timescales exceeding 10,000 years. Typically, these models include a detailed two- or three-dimensional ocean model coupled to simplified one- or two-dimensional atmospheric models. The energy balance atmospheric model predicts the geographic distribution of surface temperature and other energetic atmospheric quantities. The ocean model includes a marine biogeochemical component that simulates the chemical state of the ocean. The horizontal spatial resolution of these models is ~500 km. These models include detailed physical and biogeochemical processes that are often missing in the more complex models. However, their major limitation is that the atmospheric components are highly tuned to the present-day world, and they cannot incorporate realistic mechanical and thermodynamic forcing of the atmosphere on the ocean. Overall, these models are of value to look at transient climate change, such as the long-term fate of pCO_2 over a few hundred thousand years.

Global climate models (GCMs) are the most comprehensive models for studying the climate system (IPCC, Fourth Assessment Report, Chapter 8, 2007). These models are usually composed of three-dimensional representations of the atmosphere, ocean, sea ice, and land systems. These systems are dynamically coupled and allow for feedbacks among the various components. Recently, these fully coupled Earth System Models have begun to include other processes such as atmospheric chemistry, terrestrial and marine biogeochemistry, and ecological models. With recent increases in computational power, atmospheric GCMs are now simulating the climate system on spatial scales of 50 to 75 km (Kim et al., 2008). Lower resolution versions of fully coupled GCMs (spatial resolutions of ~100 to 400 km) can be run for thousands of years (Liu et al., 2010), and this is especially important for deep-time climate research since the equilibrium time for the oceans is ~3,000 years (e.g., Kiehl and Shields, 2005). With continued

continued

BOX 4.2 Continued

advances in computing as well as increases in the availability of massively parallel supercomputers, it is likely that multimillennial simulations using coarser-resolution GCMs (~400-km grid spacing) will soon be possible.

Such a hierarchy of climate models is essential for studying climate change on a wide range of timescales from decades to millions of years. Information from the more computationally expensive GCMs can be used as input for the EMICs, which can then be run for hundreds of thousands of years. Information from the GCMs or EMICs can also be used to better represent climate feedbacks in geochemical box models.

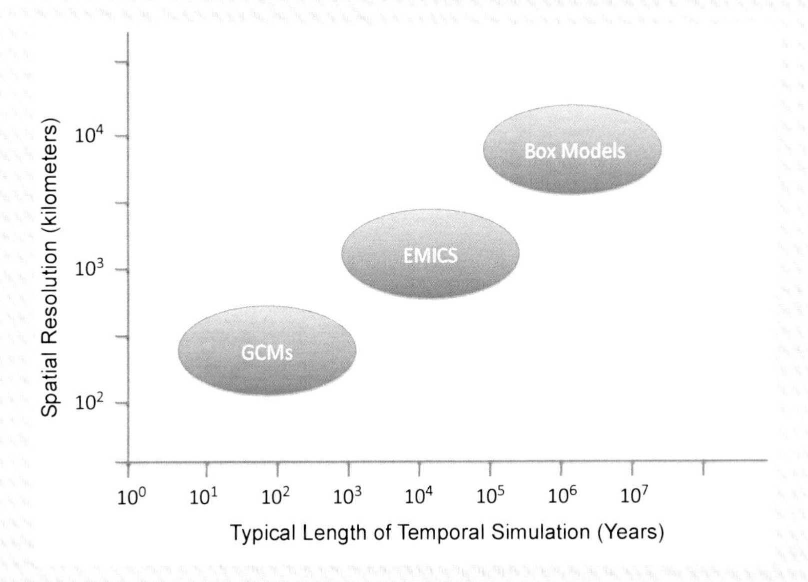

FIGURE 4.3 Graphical representation of the ranges of spatial resolution and simulation run times for the major categories of climate model.

such spatial resolutions (e.g., ~50 km). Additional challenges exist for the modeling of ancient climates, which are characterized by different paleogeography, paleotopography, atmospheric pCO_2, solar luminosity levels, and other boundary conditions. Yet, the community's confidence in the ability of GCMs to forecast future regional and global climate changes may be unfounded if these models cannot simulate past climates that differ as much from the present as the future is likely to differ from current conditions.

At this time, climate models of past periods use certain parameterizations defined for the present-day global climate system, necessarily requiring questionable assumptions about the relevance of present-day conditions to the older parts of the geological record. For example, assumptions concerning plant physiology, biome composition, and surface distribution based on present-day vegetation are not likely to have been applicable to times prior to the evolution of angiosperms (~130 million years ago [Ma]) and the expansion of grasses (~34 Ma). Similarly, the emission of precursor gases that form atmospheric aerosols is linked to current understanding of atmospheric chemistry because proxies for paleoemissions of gases that could create aerosols, in particular for those that can affect cloud properties such as biologically mediated dimethyl sulfide (Henriksson et al., 2000; Kump and Pollard, 2008), do not exist at this time. Thus, the definition of boundary conditions inferred from the geological proxy record and the parameterization of physical processes in warmer world models intrinsically come with significant uncertainty.

Key boundary conditions that must be prescribed for deep-time or warm world models include the paleogeography of land areas, past vegetation distributions, paleotopography, and ice sheet extent. Coupling to ocean models additionally requires knowledge of paleoeustasy in order to specify the distribution of shallow seas and the paleobathymetry for the deep oceans:

- Global paleogeography is an important boundary condition for constraining elevation models of continental topography and oceanic bathymetry, the geography of oceanic gateways and shallow continental (epeiric) seas that influence oceanic circulation, ocean heat transport, and climatic conditions (e.g., Crowley and Burke, 1998). The relative positions of the continents are well known back to ~200 to 180 Ma, when the oldest extant ocean floor was formed, but the uncertainty of deeper time paleogeographic reconstructions increases dramatically going farther back in time because of the absence of a seafloor record (Ziegler et al., 1983; Scotese, 2004; Blakey, 2008). Paleolatitudes can often be resolved to within about ±5°, which is somewhat coarser than the geographic resolution of the global climate models used for recent IPCC simulations of future climate. However, a greater concern is that the lack of geological record of intervening ocean basins, particularly prior to the Cretaceous, means that longitudinal control is not possible.
- Paleotopography is an important boundary condition for predicting stationary wave patterns in the atmosphere, convective effects associated with uplifted regions, and their impact on the distribution of

precipitation. Promising new approaches to reconstructing paleoaltimetry have been developed in recent years. The application of the fossil leaf stomata index to paleoaltimetry is based on the established relationship between leaf stomata frequency and ambient pCO_2 and the predictable, globally conserved decrease in pCO_2 with altitude (McElwain, 2004). The uncertainty, which can reach levels as large as the potential height of the surface (Peppe et al., 2010), is determined by the uncertainty in the CO_2 concentration in air as a function of time. The oxygen isotope compositions of pedogenic minerals (Rowley and Currie, 2006; Forest, 2007; Sahagian and Proussevitch, 2007) and the hydrogen isotope composition of *n*-alkanes from epicuticular plant waxes preserved in lacustrine deposits (Polissar et al., 2009) may be sensitive proxies of surface water and precipitation compositions and, in turn, paleoelevation through isotope-altitude relationships. As this proxy is based on systematic trends in the distribution and isotopic composition of modern precipitation with climate and topography, the uncertainty in estimates is dependent on the degree to which the isotopic composition of paleoprecipitation is faithfully recorded by the authigenic (formed in situ) minerals (Blisniuk and Stern, 2005). The recently developed "clumped-isotope" carbonate paleothermometer (see discussion below) shows good potential for paleoelevation reconstruction, using assumed temperature lapse rates with elevation (Ghosh et al., 2006; Quade et al., 2007). Recent studies have demonstrated that surface uplift influences the regional climate and isotope distribution and thereby severely complicates paleoaltimetry interpretations (e.g., Ehlers and Poulsen, 2009; Poulsen et al., 2010). Therefore, ultimately, the accurate reconstruction of paleotopography requires some degree of iteration between modeling and proxy methods since topographic relief strongly affects regional climate patterns, influencing all of the proxy-based estimates (Ehlers and Poulsen, 2009).

- Paleobathymetry has been reconstructed back to the late Jurassic through well-known age-depth relationships for oceanic crust (Parsons and Sclater, 1977; Stein and Stein, 1992), but paleobathymetry for older parts of the record is far more challenging to constrain because of the loss of seafloor through subduction. The development of oceanic plateaus and oceanic swells, and uncertainty in how the rate of ocean floor production has varied over time (Rowley, 2002; Stein and Stein, 1997), further challenges reconstructions of paleobathymetry and hence sea level change (Kominz, 1984). Besides eustasy, uncertainties in estimating changes in the rate of ocean floor production also impact estimates of mantle outgassing that have been used to drive carbon cycle models and delineate the evolution of atmospheric pCO_2, pO_2, and CH_4 through the Phanerozoic (Berner, 2006, 2009; Beerling et al., 2009). This uncertainty is significant considering that for the time intervals for which seafloor is preserved ($\leq$180 Ma), the rate of change of seafloor production remains a debated issue (Rowley, 2002).

- The Earth's vegetation contributes to, and is affected by, a variety of important climatic feedbacks and is thus an important boundary condition for paleoclimate modeling. Changes in vegetative land cover directly influence albedo and Earth surface radiation (Betts, 2000; Chase et al., 2000; Pielke et al., 2002; Marland et al., 2003; Horton et al., 2010). Changes in terrestrial vegetation also lead to changes in evapotranspiration and the hydrological cycle (Shukla and Mintz, 1982; Rind et al., 1990; Baldocchi et al., 2000; Alpert et al., 2006), with attendant indirect effects on radiative fluxes and atmospheric chemistry as a function of changes in cloud cover and water vapor mass in the atmosphere (Elliott and Angell, 1997; Hennessy et al., 1997). In addition, vegetative land cover both influences and is influenced by soil moisture content, and changes in soil moisture content can have significant climatic effects through shifts in the relative influences of latent heat flux versus sensible heating (Alpert et al., 2006; Niyogi and Xue, 2006). Vegetation-climate feedbacks have not yet been fully incorporated into GCMs because of the difficulty of parameterizing the complex, nonlinear interactions that range from cellular scale in physiological approaches to regional or global scale in biome and physical approaches to plant definition (Alpert et al., 2006). GCMs incorporating vegetation-climate feedbacks, however, generally yield amplified climate responses such as higher climate sensitivities (up to 5.5°C per CO_2 doubling; Cox et al., 2000), larger amplitudes of paleoglacioeustasy (Horton et al., 2010), and greater high-latitude amplification of warming (DeConto et al., 1999) relative to models lacking such feedbacks. The deep-time geological record offers several large-scale "natural experiments" (see Chapter 2) from which insights regarding the operation and scaling of these vegetation-climate feedbacks can be gleaned (Peteet, 2000; Parmesan and Yohe, 2003; Horton et al., 2010). The knowledge of the composition and spatial distribution of vegetation on a global scale, prior to the evolution of angiosperms (Early Cretaceous) and grasslands (Cenozoic), however, must be further developed. A far more coordinated effort is needed to expand scientific understanding of fossil plant physiological mechanisms and to synthesize disparate paleobotanical data into more comprehensive and temporally constrained compilations that can be used to refine dynamic vegetation models for climate modeling.

Further improvements in scientific knowledge of these physical and ecological boundary conditions will require more detailed analysis of paleomagnetic, paleoclimatic, paleotectonic, and paleontologic data (Van der Voo, 1993; Parrish, 1998; Kiessling et al., 1999), as well as development of more sophisticated geodynamic models. For example, there is growing evidence for a systematic shallow bias (5-10°) in paleomagnetic data from the sedimentary record (Tauxe and Kent, 2004), increasing the paleolatitudinal uncertainty of reconstructions developed using these data. Recent mantle flow simulations suggest that estimates of long-term (10 to 100 million years) eustatic sea level changes and the extent of continental flooding based on seismic and backstripping stratigraphic analysis of continental margin successions may reveal substantial mantle flow-induced dynamic topography on passive margins (Moucha et al., 2008). Furthermore, high-precision geochronological data from depositional and igneous systems worldwide are critical to constrain the ages of key paleogeographic events.

In addition to the need to better constrain the physical boundary conditions for global climate models applied to deep-time climates, the concentration of greenhouse gases (in particular CO_2) in ancient atmospheres and the solar and spectral irradiance must be determined given their fundamental contribution to radiative forcing of the climate system. The current range of uncertainty in atmospheric CO_2 and the near-complete lack of knowledge of other greenhouse gases (e.g., atmospheric methane) for many geological time periods lead to significant uncertainty in the radiative forcing of the climate system for these time periods. Improved proxies for atmospheric greenhouse gases are needed to narrow the uncertainty in the radiative forcing of the climate system.

The evolution of the solar irradiance is well constrained by solar theory over timescales of millions of years. However, variations of total solar irradiance are uncertain on timescales ranging from multidecades to multimillennia. The evolution of spectral irradiance, however, is not well constrained over geological time. Changes in spectral radiance affect the vertical deposition of shortwave energy in the atmosphere and the chemical composition of the atmosphere through photolysis processes (e.g., the formation and destruction of ozone). At present, there is no method to determine how the spectral distribution of solar irradiance has changed in the past. Finally, the latitudinal distribution of solar radiation is determined by sun-Earth geometry, and detailed celestial mechanical calculations for the temporal change of obliquity, eccentricity, and precession are limited. Clear signatures of Milankovitch cycles appear in deep-time records, but the chaotic behavior of the solar system means that Earth's orbital variations can be computed with precision for only about the past 40 million years (Laskar et al., 2004; Pälike et al., 2004).

INDICATORS OF CLIMATE SENSITIVITY THROUGH TIME—PROXIES FOR CO_2 AND MARINE TEMPERATURES

Atmospheric CO_2 Proxies

Although the physical radiative forcings from greenhouse gases are well known, anticipating the consequential rise in temperature is a much more difficult challenge because of the temporal hierarchy and complexity of feedbacks (Figure 4.4) that are triggered by CO_2 perturbation (Hansen et al., 2008). Although climate sensitivities incorporate a number of major feedbacks, reconstructions of global warming during past periods of CO_2 release indicate that there are additional short- and long-term feedbacks influencing temperature increases during CO_2-forced climate change (e.g., Zeebe et al., 2009). Significant uncertainty accompanies future climate sensitivities (~1.5 to 6°C per doubling of CO_2), with the possibility that current projections underestimate future temperatures. Calibration of climate sensitivity at "deep-time" temporal scales, which account for climate conditions that ultimately are anticipated to result from anthropogenic CO_2 release and for longer-term feedbacks that apply in warmer worlds, will be highly applicable for improved climate projections for the latter half of the twenty-first century and beyond.

Climate sensitivities on deep timescales have been estimated based on paired data for sea surface temperatures (SSTs) as well as model and proxy-based CO_2 reconstructions (e.g., Pagani et al., 2006; Royer et al., 2007; Zachos et al., 2008). For example, climate sensitivity for the transient Paleocene-Eocene Thermal Maximum (PETM) warming has been constrained using paleo-CO_2 and paleotemperature estimates inferred from stable isotopic values of marine carbonates and biomarkers—these place the climate sensitivity during the greenhouse gas-forced event between a lower bound of 2.4-3.0°C and as high as ~4°C (Pagani et al., 2006; Panchuk et al., 2008).

Estimates of paleoatmospheric pCO_2 beyond the range of ice cores are based on geochemical carbon models and an evolving toolbox of fossil organic and mineral matter proxies. Each of these methods is characterized by unique strengths, yet is limited by uncertainties and sensitivities reflecting the assumptions underlying their approach (summarized in

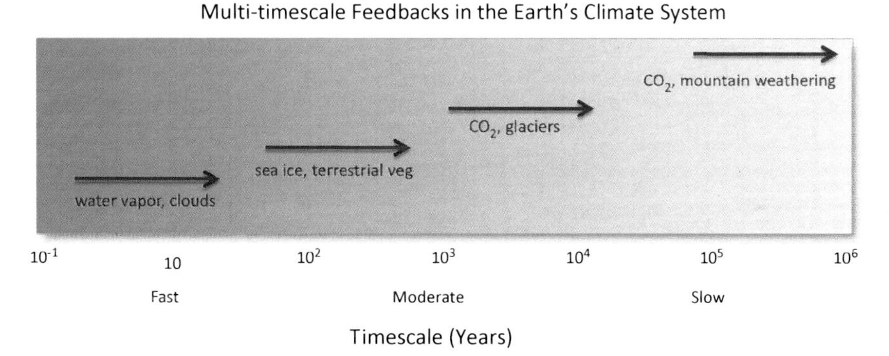

FIGURE 4.4 Schematic representation of the range of feedback scales that act on Earth's climate system.

Royer et al., 2001a; also see Box 4.3 and Table 4.1). Geochemical mass balance models (e.g., GEOCARB and GEOCARBSULF; see Berner and Kothavala, 2001; Berner, 2006) that quantify carbon transfer rates within the long-term carbon cycle have long provided a first-order assessment of how paleo-pCO_2 (and O_2) evolved throughout the Phanerozoic; many intervals have been tested and refined by comparison to proxy-based CO_2 estimates of higher temporal resolution (e.g., Ekart et al., 1999; Pagani et al., 2005; Montañez et al., 2007; Barclay et al., 2010).

The highest-precision paleoatmospheric pCO_2 estimates, in particular for the Cenozoic and Cretaceous, are based on proxy methods that utilize fossil marine and terrestrial photosynthetic flora:

- The alkenone paleobarometer, which uses the $\delta^{13}C$ of lipid biomarkers derived from haptophyte algae in marine sediments (Freeman and Hayes, 1992; Pagani et al., 1999), and
- The stomatal-index method to estimate paleo-pCO_2 from fossil leaf stomata (Woodward, 1987; McElwain and Chaloner, 1995).

Both of these proxy methods yield pCO_2 estimates with low uncertainties (tens to <100 ppmv), although marine proxy-based estimates tend to be toward the lower end of the range, while estimates inferred using terrestrial plants are generally higher (see Box 4.3). The sensitivity of these proxy methods, however, decreases dramatically above ~1,000 ppmv (±200 ppmv) due to CO_2 saturation (Bidigare et al., 1997; McElwain and Howarth, 2009).

These fossil proxy approaches, however, are based on calibrations using extant taxa or their nearest living relatives, and require the assumption that modern organisms can be used to represent biological responses of extinct organisms to ancient environments—an assumption that has been challenged. For the alkenone paleobarometer, interpretation of the photosynthetic carbon isotope effect assumes constancy in the size of phytoplankton haptophyte cells, which may have varied on millennial timescales (Henderiks and Pagani, 2007). In the case of fossil vascular plants in greenhouse periods, it is unclear how taxonomic differences and greater stomatal conductance induced by elevated temperatures (Helliker and Richter, 2008) would impact their regulation of gas exchange (Franks and Beerling, 2009). Most recently, it has been suggested that bryophytes, which lack a carbon-concentrating mechanism, may be a more reliable pCO_2 proxy (Fletcher et al., 2005, 2008), with applications to the Pleistocene (White et al., 1994) and much deeper time (Fletcher et al., 2008), yielding CO_2 estimates that are in accord with other proxy estimates of pCO_2.

Independent, mineral-based pCO_2 values, to which organic proxy and geochemical model estimates can be compared (Box 4.3), have been

BOX 4.3
Miocene Climate Change—CO_2 or Ocean Trigger?

The Miocene was a pivotal time in the current icehouse, marking the transition between fluctuating glacial conditions and rapid expansion of the Antarctic ice sheet under a cooler climate (Zachos et al., 2001a; Shevenell et al., 2004). Variations in oceanic heat transport and atmospheric vapor transport, governed by changes in deep-ocean circulation and oceanic gateways as well as decreasing pCO_2, have been implicated as drivers of this late Cenozoic global cooling (Raymo, 1994; Flower and Kennett, 1995; Holbourn et al., 2007). Recent coupled ice sheet-climate model simulations suggest that the Miocene climate transition was largely CO_2-induced, involving the crossing of a threshold pCO_2 of 400 parts per million by volume (ppmv) (Langebroek et al., 2008). Overall, estimates of pCO_2 during the Miocene range from 140 to more than 700 ppmv (Table 4.1), illustrating the critical need for improved CO_2 proxy methods (Tong et al., 2009). Moreover, contrasting estimates of atmospheric CO_2 exist for the transient period of mid-Miocene warmth (~17 Ma; compare marine-based estimates of Pagani et al. [2005], and vascular plant-based estimates of Kürschner et al. [2008]) confound the precise relationship between temperature and CO_2.

TABLE 4.1 Estimates of Atmospheric CO_2 Levels for the Middle Miocene

Reference	CO_2 (ppmv)	Method	Uncertainties
Pearson and Palmer (2000)	140 to 300	Marine $\delta^{11}B$	~20% at low CO_2, very high at >500 ppmv
Pagani et al. (1999)	180 to 290	Marine $\delta^{13}C$ (alkenone carbonate); marine $\delta^{18}O$	~20% near modern ocean conditions, infinite at $CO_2 \geq 2{,}000$ ppmv
Royer et al. (2001b)	300 to 450	Leaf stomatal indices	~20% at low CO_2, very high at >500 ppmv
Kürschner et al. (2008)	300 to 600	Stomatal frequency data from tree species	~20% at low CO_2, very high at >500 ppmv
Cerling (1991)	<700	Paleosol carbonate $\delta^{13}C$	50-100%

SOURCE: Modified from Tong et al. (2009); uncertainty estimates based on Hansen et al. (2008); Pagani et al. (2005); Royer et al. (2001b); and Cerling (1991).

calculated for much of the past 400 million years. For the Cenozoic, boron isotopes of foraminiferal calcite (assumed to mirror that of $\delta^{11}B$ of borate in seawater) have been used as a proxy for seawater pH and pCO_2 (e.g., Sanyal et al., 1997; Spivack and You, 1997; Foster, 2008; Pearson et al., 2009). Where the $\delta^{11}B$ proxy has been tested against alkenone-based pCO_2, there is a good agreement (e.g., Seki et al., 2010). In contrast, estimates of Mesozoic and Paleozoic pCO_2 are based largely on soil carbonate and goethite paleo-CO_2 proxies. The carbonate-based paleo-pCO_2 barometer (Cerling, 1991, 1992), however, is particularly sensitive to soil CO_2, a parameter that is challenging to constrain in fossil soils (Ekart et al., 1999; Breecker et al., 2009) and varies with soil moisture and productivity. Soil-formed iron oxides (goethite, gibbsite) provide a complementary mineral paleobarometer, because they form in soils that do not accumulate carbonate and the carbon isotope composition and content in their ferric carbonate component ($Fe(CO_3)OH$) are typically diagenetically robust (Yapp and Poths, 1992; Schroeder and Melear, 1999;Yapp, 2004; Tabor and Yapp, 2005). Despite the large uncertainties associated with mineral-based pCO_2 estimates, comparison of mineral- and plant-based estimates are important because of their complementary differences in sensitivity—plant-based proxies lose sensitivity above 800 to 1,000 ppmv, whereas mineral-based proxies are more sensitive above 1,000 ppmv (Royer et al., 2001a)—and the lack of extant plant calibrations for stomatal index-based estimates in pre-Cretaceous intervals.

Marine Temperatures

A major challenge in determining ancient climate sensitivity is the need for robust estimates of ocean temperatures and latitudinal ocean temperature gradients. This need is particularly great for climate reconstructions of the pre-Cretaceous, for which we presently have only binary "icehouse or greenhouse" reconstructions (Figure 1.2) and highly interpretive and largely unpublished climate syntheses (e.g., PALEOMAP Project[1]). Several proxies have been employed to reconstruct ancient sea surface and deep-water temperatures, each with its strengths and limitations. Measuring multiple proxies not only provides refined SST reconstructions but also offers constraints on taxonomic (vital), environmental, and diagenetic influences that might affect each proxy. For several decades, the gold standard for reconstructing sea surface and bottom temperatures for the post-Jurassic has been from $\delta^{18}O$ values of foraminifera (e.g., Shackleton, 1987; Zachos et al., 2001a), whereas for older geological intervals, it has been the $\delta^{18}O$ values of metazoan skeletal calcites (Veizer

[1] See http://www.scotese.com/climate.htm.

et al., 1999; Grossman et al., 2008) and conodont apatites (Joachimski et al., 2006; Buggisch et al., 2008). There is, however, growing appreciation for the susceptibility of fossil carbonate to diagenesis and thus overprinting of the precipitation (seawater) temperature signal in its $\delta^{18}O$ values (Box 4.4) (Schrag et al., 1995; Pearson et al., 2001, 2007; Came et al., 2007; Grossman et al., 2008; Kozdon et al., 2009; Cochran et al., 2010). Furthermore, carbonate $\delta^{18}O$ values are influenced by both temperature and seawater $\delta^{18}O$, which in turn has varied through time due to varying ice sheet $\delta^{18}O$ composition, their effects of waxing and waning ice sheets on seawater $\delta^{18}O$, and surface net evaporation balance, in particular in the broad, shallow epicontinental seas of the pre-Cenozoic (Holmden et al., 1998; Veizer et al., 1999; Panchuk et al., 2006). The compound influence of changing seawater $\delta^{18}O$ and temperature on carbonate $\delta^{18}O$ values for periods that were not ice-free and the possible effects of seawater alkalinity on carbonate $\delta^{18}O$ (Spero et al., 1997; Zeebe, 1999) have led to increased focus on independent temperature proxy methods.

The Mg/Ca ratios of planktonic foraminiferal calcite have been shown to be sensitive to temperature, increasing exponentially with increasing SSTs, providing an SST paleothermometer (e.g., Elderfield and Ganssen, 2000; Lear et al., 2000). Application of the Mg/Ca temperature proxy to deep-sea sediments has challenged climate change paradigms by documenting dynamic glacial-interglacial temperature variation in Pleistocene tropical oceans and linkages to extratropical climates, and refining phasing between changes in atmospheric CO_2, surface and deep-ocean temperature changes, and Antarctic glaciation (Lea et al., 2000; Zachos et al., 2008). Mg/Ca paleothermometry, however, requires species-specific calibration to account for environmental and vital (taxonomic) effects and may be limited by salinity, pH, and/or carbonate ion effects on magnesium partitioning in foraminiferal calcite (Dekens et al., 2002, 2008; Lear et al., 2002, 2008; Russell et al., 2004; Coxall et al., 2005; Ferguson et al., 2008; Hoogakker et al., 2009). Importantly, recent multiproxy reconstructions of deep-sea sediments document good agreement between alkenone-based and Mg/Ca temperature estimates (Bard, 2001; Dekens et al., 2008). In addition to deconvolving the temperature and ice volume signals in the isotopic record, integrating foraminiferal $\delta^{18}O$ values and Mg/Ca ratios has permitted the reconstruction of oceanic surface salinity distributions and their effect on oceanic heat transport (e.g., Schmidt et al., 2004). Efforts to apply the Mg/Ca temperature proxy to pre-Cenozoic biotic calcites (e.g., mollusks) have been limited by the influence of vital effects in these calcifying organisms (Immenhauser et al., 2005), variation in seawater alkalinity and $[CO_3^{2-}]$, and the need to account for likely secular variation in seawater Mg/Ca and ocean saturation state.

BOX 4.4
Proxies and the "Tropical Climate Paradox"

Proxy evidence for mid- to high-latitude warmth in both the marine (Crowley and Zachos, 2000; Zachos et al., 2001a) and the continental (Greenwood and Wing, 1995; Wolfe, 1995; Markwick, 1998) realms of the Early Eocene (57-50 Ma) are consistent with global warmth and a high-CO_2 atmosphere (Pearson and Palmer, 2000; Pagani et al., 2005). However, marine isotope data from low paleolatitudes initially indicated little or no ocean surface warming, and perhaps even cooling, in the Eocene tropics (Shackleton and Boersma, 1981; Barron, 1987), leading to what was termed the "cool tropics paradox" (Barron, 1987). Subsequently, scientists have revisited the low-latitude planktonic foraminiferal record (Pearson et al., 2001, 2007; Norris et al., 2002; Kozdon et al., 2009), arguing that diagenesis on the cold deep seafloor has imparted a cool overprint signal to the oxygen isotopic composition of warm-water planktonic foraminifera, biasing the record in the tropics toward cold temperatures.

The tropical climate paradox may be fully resolved with new estimates of tropical SST for the Eocene derived from pristine foraminifera and a new organic molecule proxy, TEX_{86},[a] that indicate temperatures 5-10°C warmer than previous reconstructions (Pearson et al., 2007) (Figure 4.5). Such paleotemperature reconstructions further challenge the tropical thermostatic regulation hypothesis presented in Chapter 2 and imply that the cooling trend of the Eocene was primarily a high-latitude phenomenon with little effect on the tropics, where climate remained warm and stable (Pearson et al., 2007).

[a] Tetraether index of 86 carbon atoms; paleothermometer based on the composition of membrane lipids of marine picoplankton.

Several additional proxy methods show promise for paleothermometry but are still in the development and testing phases:

- The calcium isotope ($\delta^{44}Ca$) composition of well-preserved foraminifera may provide an independent paleotemperature proxy and a test of the reliability of the Mg/Ca proxy (e.g., Nägler et al., 2000), although the complex calcium isotope fractionation behavior, ancient seawater $\delta^{44}Ca$, and possible vital effects during biocalcification are not yet fully understood (Gussone et al., 2009). Limited $\delta^{44}Ca$ data for Cretaceous rudist

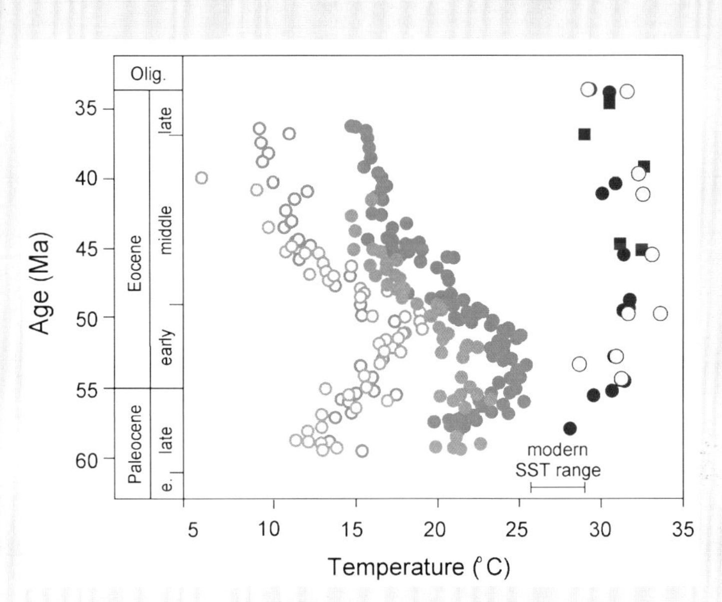

FIGURE 4.5 Tropical paleotemperature estimates based on $\delta^{18}O$ data from unaltered benthic and planktonic foraminifera (red circles and squares, respectively) and from TEX_{86} data (yellow circles) from onshore drill core samples from Tanzania. These are plotted for comparison with benthic (open circles) and planktonic (green circles) foraminifera $\delta^{18}O$ data from ODP cores collected from the tropical Pacific Ocean, inferred to have been diagenetically altered and thereby indicating anomalously low paleotemperature estimates.
SOURCE: Pearson et al. (2007).

calcite demonstrate the potential for extending this temperature proxy to pre-Cenozoic calcitic macrofauna (Immenhauser et al., 2005). Similarly, the magnesium isotope ($\delta^{26}Mg$) composition of aragonite corals shows promise as a seawater paleothermometer (Wang et al., 2008).

- Recent studies have documented that the clumping (ordering) of carbon and oxygen isotopes into bonds in biogenic and abiotic carbonates is temperature dependent and an independent record of the fluid $\delta^{18}O$ in which the carbonate precipitated or stabilized during diagenesis ("clumped isotope method"; Came et al., 2007; Eiler, 2007; Tripati et al.,

2010). The carbonate clumped isotope proxy holds considerable promise for constraining marine and continental (e.g., speleothems, vertebrate bioapatite) paleotemperatures, although kinetic and diagenetic effects must be better understood (Affek et al., 2008; Came et al., 2008; Eagle et al., 2010).

- Biomarker proxies permit reconstruction of paleo-SSTs that are independent of mineral-based proxy estimates, and include $U^{K}_{37'}$ (Brassell et al., 1986), which is based on the relative abundances of C_{37} alkenones photosynthesized by marine green algae, and the more novel biomarker, $TEX_{86'}$ which is based on the relative abundances of C_{86} tetraether lipids that form in the water column by archaeal microbiota (Schouten et al., 2002; Eglinton and Eglinton, 2008). The low diagenetic susceptibility of these biomarkers and the preservation of tetraether lipids in sediments as old as the Cretaceous have provided new insight into the climate dynamics of the recent icehouse (Haug et al., 2004; Kienast et al., 2006; Martrat et al., 2007; Dekens et al., 2008) through Cretaceous and Eocene greenhouse periods, including contributing to the resolution of the Cretaceous cool tropics paradox (see Box 4.4) (Pearson et al., 2007; Schouten et al., 2007). Integration of multiple proxies greatly increases the range of paleotemperature sensitivity and probably also increases the precision of estimates because of the variable sensitivity of different proxies. For example, the Mg/Ca proxy is least sensitive at low temperatures, whereas the $U^{k'}_{37'}$ method is least sensitive at high temperatures.

INDICATORS OF REGIONAL CLIMATES

Climate models and paleoclimate archives indicate that one of the larger impacts of global warming is likely to be regional changes in continental temperatures and precipitation. It is thus imperative to constrain past temporal and geographic variability in continental climate—in particular for periods of abrupt and/or major climate transitions and for climates that were warmer than the present day—in order to better understand how regional climates may change in the future. For continental settings, deep-time paleoclimate reconstructions require a multiproxy approach involving comparable proxies.

Estimating Continental Temperatures

There are numerous and reasonably well-developed proxies for estimating continental paleotemperatures from lacustrine, coastal, and terrestrial deposits. Fossil plant leaves and pollen have been the major source of continental paleotemperature estimates, because the composition and physiological properties of plant communities change rapidly with temperature (and relative humidity) changes. For flowering plants (angiosperms), the shapes (style of leaf margin) and sizes (physiognomy) of fossil leaves have been shown to vary with mean annual temperature (MAT) (Wolfe, 1993; Wilf, 1997; Kowalski and Dilcher, 2003; Royer et al., 2005) and have been applied throughout the Cenozoic, including for the PETM (Wing et al., 2005). Plant-based continental temperature estimates for periods that predate the evolution of angiosperms (pre-Cretaceous), however, are missing. Estimates of paleocontinental temperatures also have been interpreted from pollen distributions in lake sediments, through comparison with the temperature sensitivity of modern plant communities (Overpeck et al., 1985). The pollen distribution approach, however, becomes significantly less certain on longer timescales due to increased differences among ancient and modern plant species and communities.

Mineral-based isotopic paleothermometry offers an independent set of proxies that are not restricted stratigraphically to the post-Jurassic, and for which interpretations are not limited by lack of calibration to extant plants. For example, the $\delta^{18}O$ values of pedogenic carbonates and $\delta^{18}O$ and δD values of hydroxylated clay minerals (kaolinite and smectite) and iron oxides (goethite and hematite) from fossil soils are being used to estimate paleotemperatures in ancient soils hundreds of millions of years old (Dworkin et al., 2005; Tabor and Montañez, 2005), although this approach requires assumptions regarding the stable isotope composition of meteoric water. This limitation can be overcome for soil carbonates by application of the carbonate clumped isotope thermometer to paleosol carbonates (Passey et al., 2010) and through oxygen isotope–mineral pair thermometry between coexisting pedogenic clays and iron oxides (Tabor, 2007). Both methods allow for paleosoil temperature estimates that are independent of the soil water $\delta^{18}O$ in which the minerals formed. Conventional $\delta^{18}O$ and clumped isotope analysis of vertebrate bioapatites (and $\delta^{18}O$ analysis of body scales of some freshwater fish) offer yet another independent proxy of continental MAT and have been used to reconstruct MAT geographic patterns for deep-time warm periods (e.g., Koch et al., 2003; Fricke and Wing, 2004) and to constrain body temperatures of extant and extinct vertebrates (e.g., Barrick and Kohn, 2001; Eagle et al., 2010).

In addition to isotopic paleothermometry, the major element compositions of paleosols have been used for MAT reconstructions as far back as the Paleozoic, based on transfer functions calibrated using modern soils and associated mean annual temperatures (Sheldon et al., 2002; Retallack, 2005). Applications of this proxy yield estimates of Cenozoic paleo-MAT that are consistent with fossil leaf morphology-based paleotemperature estimates (Sheldon, 2009). For all of these mineral-based proxies, the accuracy and uncertainty of paleotemperature estimates are largely dependent on using appropriate fossil material that has been minimally altered by diagenesis.

Ultimately, reconstructions of regional variation in continental temperatures will require further calibration studies of existing proxy methods along with the development of new higher-resolution proxies and continued development of spatially highly resolved multiproxy datasets. One promising new direction is the use of biomarkers preserved in lacustrine and marginal marine sediments that appear to be diagenetically relatively robust. The presence of picoplankton members of the Archaea in ancient lake deposits opens up the possibility of using the relative abundances of their membrane lipids (TEX_{86}) as a paleothermometer of surface water (Eglinton and Eglinton, 2008). Microbial lipid patterns in modern soils also show potential as a biomarker paleotemperature proxy, but the influence of other soil parameters on their abundances (e.g., pH) requires further calibration studies.

Estimating Regional Hydroclimates

The understanding of regional patterns for continental paleoclimate changes in the younger part of the record comes primarily from a wealth of paleolacustrine records and a rapidly increasing speleothem paleoclimate database. Lacustrine records offer the continuity and temporal resolution potential to provide key high-resolution sedimentological, geochemical, and paleontologic time series for reconstructing paleocontinental regional climate change. The paleo-water balance of some ancient lakes, such as those in the western United States, also provides strong signals of regional changes in effective moisture (e.g., Benson et al., 2003). Speleothem physical and geochemical proxies are proving to be powerful recorders of changes in regional air temperature and effective moisture (e.g., Oster et al., 2009; Wagner et al., 2010). Notably, precisely dated (U-series) speleothem records are revealing rapid—century to subdecadal—climate anomalies, often involving inter- and intrahemispheric atmospheric teleconnections (Wang et al., 2001, 2005; Yuan et al., 2004). However, the bulk of these continental records are from the late Cenozoic, and thus have not been used to reconstruct regional hydroclimate patterns during warm Earth climates.

Reconstructing regional patterns in relative humidity and precipitation is far more challenging in deep-time records because of the overall lower levels of temporal and spatial resolution, stratigraphic continuity, and geochemical susceptibility to diagenesis, although several new approaches are being evaluated. For much of the pre-Neogene, scientific understanding of climate regimes is based on low spatial and temporal resolution global syntheses of published databases (Ziegler et al., 2003; Boucot et al., 2004). The morphological characteristics of ancient soils and their bulk geochemical composition have climatic significance because the intensity of pedogenesis is dominantly related to precipitation patterns and surface temperature. Quantitative proxies for estimating mean annual precipitation have been developed that use the iron content in pedogenic Fe-Mn nodules, the depth to the pedogenic carbonate horizon, or the chemical composition of particular paleosol horizons (known as the Chemical Index of Alteration, CIA), all of which are based on empirical relationships derived from modern soils (Stiles et al., 2001; Sheldon et al., 2002; Retallack, 2005; Sheldon and Tabor, 2009). These approaches have been applied to a wide age range of Phanerozoic paleosols (e.g., Driese et al., 2005; Prochnow et al., 2006), including to the PETM where these soil proxy-based precipitation estimates indicate a transient drying in western North America associated with transient global warming (Kraus and Riggins, 2007). Where multiproxy records permit, CIA-based estimates of mean annual precipitation are consistent with independent paleobotanical estimates. The measured $\delta^{18}O$ compositions of ancient soil-formed minerals (phyllosilicates, carbonates, iron oxides, and sphaerosiderites) have been shown to be reliable proxies of soil-water $\delta^{18}O$ and, in turn, $\delta^{18}O$ precipitation at a given paleolatitude after consideration of evaporative or nonequilibrium effects (Stern et al., 1997; Yapp, 2000; Vitali et al., 2002; Ufnar et al., 2004; Tabor and Montañez, 2005). The fact that these minerals form in equilibrium with ambient hydrological conditions means that they provide a sensitive record, where formed, of shifts in seasonality and precipitation rates, allowing them to be used to evaluate the hydrological cycle in past greenhouse worlds and periods of icehouse-to-greenhouse transition.

In the same way that marine and lacustrine biomarkers have been used as quantitative paleothermometers, the hydrogen isotope ratios (δD) of individual lipids in fossil plant tissues show great potential for reconstructing paleocontinental hydrological conditions. Leaf wax *n*-alkanes are some of the most abundant lipid molecules biosynthesized by terrestrial plants (Eglinton and Hamilton, 1967), containing C-bound hydrogen that is geologically stable (Schimmelmann et al., 1999, 2006). Plant *n*-alkane δD values have been shown to correlate with local meteoric water δD (Sternberg, 1988), further modified by isotope enrichment in leaf water via transpiration and soil water evaporation (Sachse et al., 2006). Given the control of relative humidity on these processes, fossil leaf wax *n*-alkanes are being explored as a paleoaridity proxy (e.g., Liu and Huang, 2005; Pagani et al., 2006; Smith and Freeman, 2006). Scientific understanding of the role of climate and plant physiology on compound-specific δD systematics is still evolving (Chikaraishi and Naraoka, 2003; Pedentchouk et al., 2008; Diefendorf et al., 2010), requiring further empirical study of plant-water-deuterium systematics before this proxy can be applied straightforwardly to ancient continental systems. Further development and refinement of emerging proxies such as the aforementioned mineral stable isotope and biomarker approaches, along with improved spatial and temporal resolution of terrestrial proxy records, are fundamental to successful calibration and testing of climate models for a future warmer Earth using deep-time analogues.

INDICATORS OF OCEANIC PH AND REDOX

Recent advances in instrumentation (e.g., multicollector inductively coupled plasma mass spectrometry, nanoscale secondary ion mass spectrometry) coupled with development and modern calibration of geochemical and isotopic proxies that span the periodic table have greatly expanded the range of paleoceanographic proxies of oceanic redox, alkalinity, and pH. Studies over the past decade have documented the linear or exponential relationships between trace element ratios (Mg/Ca, Cd/Ca, Zn/Ca, U/Ca) and stable (O, C, B) isotopic compositions of carbonate-bearing fossil fauna and changes in seawater carbonate content $[CO_3^{2-}]$ and pH (Lea et al., 1999; Marchitto et al., 2000; Russell et al., 2004). For example, the U/Ca ratios of certain planktonic foraminifera genera have been used to determine variations in seawater carbonate ion content over the last glacial cycle that track atmospheric pCO_2 variations archived in polar ice cores (Russell et al., 1996, 2004).

The timing and geographic extent of past events of oceanic hypoxia and anoxia can be resolved through the integration of sulfur isotopes of reduced (pyrite) and oxidized minerals (carbonate-associated sulfate in carbonates) and fossil organic matter, and the abundance and stable isotopic composition of heavy metals (e.g., Fe, U, and Mo isotopes). The abundance of redox-sensitive transition elements (V, Mo, Fe, Cr) and their partitioning between various mineral phases in organic-rich deposits have provided much insight into the origin of O_2-deficient waters in pre-Cenozoic marine basins—in particular in past greenhouse worlds (e.g., Sageman et al., 2003; Meyers et al., 2005; McArthur et al., 2008b; Lyons et al., 2009; Algeo et al., 2010). Carbonate-associated sulfate data coupled with pyrite sulfur isotope data from Precambrian oceanic deposits have refined scientific understanding of the oxygenation of Earth's early atmosphere (Kah et al., 2004). Recently, the utility of molybdenum ($\delta^{97}Mo$), uranium ($\delta^{238}U$), and iron ($\delta^{56}Fe$) isotopes of various components in organic-rich black shales has been demonstrated as a sensitive proxy of oceanic O_2 levels, revealing the protracted oxygenation of Earth's early ocean in the Proterozoic (e.g., Anbar and Knoll, 2002; Arnold et al., 2004), and elucidating the global expansion of oceanic anoxia during past warm periods (e.g., Jenkyns et al., 2007; Gordon et al., 2009; Duan et al., 2010; Montoya-Pino et al., 2010).

Synergy of Observations and Models

A synergistic approach combining observations and modeling provides an optimal strategy for answering critical questions regarding how Earth's climate system has responded to varying levels of greenhouse gases and other forcing factors. Model simulations provide a global picture of the state of the climate system and also a window into how various processes operate to maintain a given climate state. Disparities between simulated climate variables (e.g., surface temperatures, precipitation, ocean circulation) and proxy observations of these state variables pose questions regarding how much of the disparity is due to model biases or deficiencies and how much is due to observational bias. An example of data bias is related to simulated tropical and subtropical SSTs during warm Paleogene climates, which for years were high compared to proxy data. Recent recognition and correction of problems with the proxy data (see Box 4.4) have not brought models and data into greater agreement. An example of model bias is related to simulated high-latitude surface temperatures in warm climate regimes, which have been too low compared to proxy data. Here, continued improvement and development of new innovative observational techniques have strengthened the conclusion that models are challenged to simulate such high polar surface temperatures. This disparity has led to active model exploration of feedback processes that may operate in warm greenhouse climates but are not revealed by data-model studies of Earth's more recent glacial state. Thus, it is to the benefit of both observational and modeling communities to work in close collaboration through real-time data-model comparison studies. Disparities between models and observations represent synergistic research opportunities.

5

Implementing a Deep-Time Climate Research Agenda

The present state of scientific knowledge regarding the deep-time record of climate change, summarized in previous chapters, highlights the insights that have been gleaned from studies of past warmings and major climate transitions, including some that are analogues for anticipated future climates. This research status outline provides an indication of the most important enduring issues that will require further research and points to the potential for dramatic scientific discovery in the largely untapped deep-time record. This chapter presents a scientific research agenda designed both to answer the series of major questions posed in Chapter 1 regarding the impact of the projected rise in atmospheric pCO_2 and to provide a more refined understanding of the important processes—uniquely present in the deep-time geological record—that will drive the Earth system as it transitions to a warmer world. The chapter also describes the research tools and community effort that will be required to implement this research agenda and provides recommendations for an education and outreach strategy designed to broaden scientific and general community understanding of the contribution that can be derived only from the deep-time record. Finally, the committee stresses the need to bolster existing mechanisms, and design new mechanisms, for bringing together interdisciplinary collaborative scientific teams from diverse fields to focus on the insights that can be gleaned from the deep-time geological record and to ensure the maximum integration and sharing of the diverse databases that will result from this research.

ELEMENTS OF A HIGH-PRIORITY DEEP-TIME CLIMATE RESEARCH AGENDA

The workshop hosted by the committee provided a wealth of information concerning the existing scientific status of deep-time climate research, as well as a very broad range of topics that the community suggested as research foci for an improved understanding of Earth system processes during the transition to a warmer world. The committee assessed these topics and their potential to transform scientific understanding, and identified the following six elements of a deep-time research agenda as having the highest priority to address enduring scientific issues and produce exciting and critically important results over the next decade or longer.

Improved Understanding of Climate Sensitivity and CO_2-Climate Coupling

Existing data indicate that climate forcing resulting from increased CO_2 will, by the end of this century, rival that experienced during past greenhouse periods prior to the onset of the current glacial state. The paleoclimate record, which captures the climate response to a full range of levels of radiative forcing, can uniquely contribute to a better understanding of how climate feedbacks—both long and short term—and the amplification of climate change have varied with changes in atmospheric CO_2 and other greenhouse gases. In the context of the large uncertainty in estimates of climate sensitivity described in Chapters 1 and 2, a high research priority for deep-time paleoclimatology is the determination of *equilibrium* climate sensitivity on multiple timescales, particularly during periods of greenhouse gas forcing comparable to that anticipated within and beyond this century if emissions are not reduced. Existing records of past warm periods already indicate climate sensitivity well above the estimated short-term range and show that the future temperature increase will most likely be amplified once the longer-term feedbacks that have not operated on human timescales (decades to centuries) during Earth's current icehouse become relevant under warmer conditions.

Further mining of the deep-time geological archive will require focused efforts to improve the accuracy and precision of existing proxies for past atmospheric pCO_2 and surface air and ocean temperatures, and to develop new proxies for other paleo-greenhouse and non-greenhouse gases and aerosols. Data using new and existing proxies could then be synthesized to develop an authoritative global temperature and atmospheric pCO_2 history—at various resolutions—for the full span of Earth's history. Improved constraints on levels of radiative forcing and equilibrium climate sensitivity are needed for past warm periods and major climate transitions. In addition, further study of intervals of possible CO_2-climate decoupling (e.g., mid-Miocene, Late Jurassic, Early Cretaceous) will require careful integration of paleoatmospheric CO_2 and paleotemperature time series with improved temporal resolution, precision, and accuracy, as well as data-model comparisons to critically evaluate the veracity of these apparent mismatches. With these improved data, a hierarchy of models can be used to test various forcing mechanisms (e.g., non-CO_2 greenhouse gases, solar, aerosols) to determine how well mechanisms other than CO_2 can explain anomalously warm and cold periods and to critically evaluate the climate processes and feedbacks that led to particular climate responses characteristic of greenhouse gas-forced climate changes in the past.

Climate Dynamics of Hot Tropics and Warm Poles

Paleoclimate observations provide a conundrum that must be resolved to understand the climate system—the evidence that past temperatures in the tropics and polar regions were periodically much hotter than today. How can the Earth maintain tropical temperatures approaching 40°C, or how can polar temperatures remain above freezing year-round? Yet there is very strong evidence for both conditions during past warm periods. The deep-time paleoclimate evidence suggests that the mechanisms and feedbacks in the modern icehouse climate system that have controlled tropical temperatures and a high pole-to-equator thermal gradient may not apply straightforwardly in warmer worlds. Moreover, the fundamental mismatch between climate model outputs, modern observations, and paleoclimate proxy records discussed in Chapter 2 highlight the degree to which science's current understanding of how tropical and higher-latitude temperatures respond to increased CO_2 forcing remains limited. An improved understanding of these processes, which may drive significant changes in surface temperatures in a future warmer world, is imperative given the potential dire effects of higher temperatures on tropical ecosystems and the domino effect of polar warming on ice sheet stability, the stability of permafrost (which carries a large load of greenhouse gases), and regional climates through atmospheric teleconnections with the tropics and/or polar regions.

Accomplishing this goal requires that the range of deep-time observational data be expanded to include latitudinal transects that span the tropics through mid- to high-latitude regions for targeted intervals of Earth history. Improved constraints on the meridional thermal structure of warm worlds will require increased chronological constraints and more spatially resolved proxy time series than currently exist. New theoretical and modeling approaches are also required to develop a comprehensive understanding of the limits of tropical and polar climate stability, and an understanding of how a weaker thermal gradient is established and maintained in warmer climate regimes. Global climate models (GCMs) offer an astounding array of diagnostics for assessing atmospheric dynamics and teleconnections, but these diagnostics need to be employed far more commonly in analyses of paleoclimate simulations, requiring deeper and "real-time" collaborations between the "observationists" and the atmospheric dynamicists. The documented ability to successfully model conditions comparable to those anticipated in the future will provide a test of the efficacy of climate model projections for continued global warming.

Sea Level and Ice Sheet Stability in a Warm World

Large uncertainties in the theoretical understanding of ice sheet dynamics and associated feedbacks currently hamper the ability to predict how the ice sheets currently in the Earth's polar regions, and sea level, will respond to continued climate forcing. For example, paleoclimate studies of intervals within the current icehouse document variability in ice sheet extent that cannot be reproduced by state-of-the-art coupled climate-ice sheet models. Moreover, studies of past warm periods indicate that equilibrium sea level in response to current warming may be substantially higher than model projections indicate due to the influence of dynamic processes that have not been operative in the recent past. Efforts to address these issues will have to focus on past periods of ice sheet collapse that accompanied transitions from icehouse to greenhouse conditions, to provide context and understanding of the "worst-case" forecasts for the future.

Future studies that probe deeper into Earth history should focus on periods that have the potential to reveal critical threshold levels associated with ice sheet collapse and to elucidate the dynamic processes and feedbacks that have led to deglaciation in the past but are not captured by paleoclimate records of the past few million years. An integral component of such studies should be a focus on improving science's ability to deconvolve the temperature and seawater signals recorded in biogenic marine proxies, including refinement of existing paleotemperature proxies and the development of new geochemical and biomarker proxies. Modeling the distribution of ice in warm worlds will need to expand beyond the intermediate-complexity models that currently include this component in order to involve the coupling of land ice component models to complex GCMs and include full interaction with the atmospheric hydrological cycle.

Understanding the Hydrology of a Hot World

Studies of past climates and climate models strongly suggest that the greatest impact of continued CO_2 forcing will be regional climate changes, with ensuing modifications to the quantity and quality of water resources—particularly in drought-prone regions—and impacts on ecosystem dynamics (Lunt et al., 2008; Haywood et al., 2009; Shukla et al., 2009). Because of the sensitivity of climate to small changes in high-latitude and tropical temperatures, an improved understanding of the hydrological cycle during periods of increased radiative forcing—comparable to those projected for the future—is imperative. Because of the potential for large feedbacks to the climate system, this in turn requires an improved understanding of the interaction between the global hydrological and carbon cycles over a full spectrum of CO_2 levels and climate conditions. The deep-time record uniquely archives the physical and geochemical expressions of the carbon and water cycle dynamics that operated during past warm periods, including the response of low-latitude precipitation to high-latitude unipolar glaciation or ice-free conditions (e.g., Floegel and Wagner, 2006; Poulsen et al., 2007a,b; Ufnar et al., 2008), the stability of continental carbon reservoirs (soils, wetlands, tundra, permafrost) to changing regional climates, and the impacts on—and response of—ecosystems to such changes.

These research objectives require the development of marine-terrestrial transects with spatially resolved proxy records at high temporal resolutions and precisions. In particular, paleoterrestrial reconstructions have long been plagued by sparse and discontinuous outcrop, stratigraphically incomplete successions, and poor chronological constraints. The optimum approach is thus to integrate chronostratigraphically well-constrained marine records with contemporaneous terrestrial records through integration of radiometric, biostratigraphic, and/or magnetostratigraphic data. The implementation of this objective will require transect-focused ocean and continental drilling.

Efforts to improve existing proxies, to develop new proxies, and to develop multiproxy time series in order to provide quantitative estimates of paleoprecipitation, paleoseasonality, paleohumidity, and paleosoil conditions (including paleoproductivity) are a critical component of this research, in particular where the level of precision—and thus the degree of uncertainty in inferred climate parameter estimates—can be significantly reduced. Proxy improvement efforts should include strategies for better constraining the paleogeographic setting of proxy records, including latitude and altitude or bathymetry.

Understanding Tipping Points and Abrupt Transitions to a Warmer World

Studies of past climates and climate models show that Earth's climate system does not respond linearly to gradual CO_2 forcing, but rather responds by abrupt change as it is driven across climatic thresholds. Modern climate is changing very rapidly, and there is a possibility that Earth will soon pass thresholds that will lead to even more rapid changes in Earth's environments. It is possible that such thresholds could involve transition into a new climate state that cannot return to pre-CO_2 forcing conditions if the prior conditions are reestablished. Thus the proximity of Earth to such a 'tipping point' is a critical question. The answer does not reside in the more recent paleoclimate record, but rather is to be found in the dynamics of past transient events where the climate system crossed critical thresholds into climate states more representative of where Earth's climate may be heading. Because of their proven potential for capturing the dynamics of past abrupt changes, intervals of rapid (millennia or less) climate transitions in the geological record—including past hyperthermals—should be the focus of future fully integrated paleoclimate, paleoecologic, and modeling collaborations. Key insights to be gleaned from such studies include an improved understanding of how various components of the climate system responded to such abrupt transitions, in particular during times when the rates of change were sufficiently large to imperil biotic diversity. There is also a need to understand where to expect thresholds and feedbacks in the climate system—especially in warm worlds and past icehouse-to-greenhouse transitions. Moreover, targeting such intervals for more detailed investigation is a critical requirement for constraining how long any such climate change might persist.

Key requirements for an improved understanding of abrupt climate change are better dynamic models and datasets to resolve the behavior of the environment in transition. On the data side, substantially improved spatial (subkilometers to tens of kilometers over large geographic regions) and temporal (subcentennial scale) resolution of datasets from Earth's past are required to illustrate the behavior of environmental systems in rapid transition. These include both examples of transition into fundamentally new climate states and examples of transient climate states that ultimately returned to near preperturbation conditions (e.g., the Paleocene-Eocene Thermal Maximum [PETM]). To be most effective, temporal resolution on the level of centuries or less is needed to identify and understand climate and ecosystem changes at rates relevant to human society.

Current predictions of the duration of future greenhouse conditions are based on simplified models of the climate system and carbon cycle, constrained by limited observations of their behavior during analogous times in Earth history. A more convincing answer to the central question of "how long" requires more sophisticated and comprehensive models, and it will be possible to have confidence in the models only if they can be evaluated against observations. Intermediate-complexity models that are capable of running continuous simulations for the 10,000 to 100,000-year duration of these events are needed, and such models need to treat the ocean and atmosphere as an open system as the basis for spatial and temporal predictions that can be directly compared with observational data of similar temporal resolution. Specifically, it is important that the models calculate variables that are similar to those measured in the field or calculated using proxy methods, so that direct comparisons between paleoecosystem proxies and model results are possible—for example, the inclusion of oxygen and carbon isotopes as tracers in both atmosphere and ocean models or, in the case of models of intermediate complexity, the inclusion of sediment transport modules. The historical record and even the broad expanse of the Pleistocene climate record contain nothing comparable to the anticipated outcomes following the burning of all fossil fuel resources, and thus cannot be considered appropriate analogues from which to refine an understanding of the climate and ecosystem changes that continued warming will cause.

Understanding Ecosystem Thresholds and Resilience in a Warming World

Both ecosystems and human society are highly sensitive to abrupt shifts in climate because such shifts may exceed organism tolerances and consequently have major effects on biotic diversity as well as human investments and societal stability. Modeling future biodiversity losses and biosphere-climate feedbacks, however, is inherently difficult because of nonlinear interactions and the existence of both positive and negative feedbacks that add complexity to the system and increase the uncertainty of the net response to climatic forcing. How rapidly biological and physical systems can adjust to abrupt climate change is a fundamental question accompanying present-day global warming. An important tool to address this question is to describe and understand the outcome of equivalent "natural experiments" in the deep-time geological record, in particular where the magnitude and/or rates of change in the global climate system were sufficiently large to threaten the viability and diversity of species, leading at times to mass extinctions. The paleontological record of the past few million years does not provide such an archive because it does not record catastrophic-scale climate and ecological events.

The deep-time record of past biotic turnovers and mass extinction events associated with warm periods (many associated with massive outgassing of CO_2 or methane), transient warmings, and major transitions between climate states offers an undertapped repository from which unique insights can be obtained regarding patterns of ecosystem stress, the potential for ecological collapse, and mechanisms of ecosystem recovery. For example, integrated paleoclimate and paleoecology studies can uniquely address the fundamental question of how hot the tropics will become, and how much ocean chemistry will be perturbed, under additional CO_2 radiative forcing. This is a critical issue because such changes may have dire effects on tropical ecosystems, with the potential for severe declines in diversity over large areas. For example, studies of past greenhouse gas-forced transient warmings provide the only analogue of the future potential for ocean acidification and its effect on calcifying organisms. The penultimate deglaciation of the Late Paleozoic Ice Age is the only archive recording how tropical floral ecosystems might respond to climate change associated with an epic deglaciation. The issue of how Arctic ecosystems will respond if sea ice disappears permanently and/or the Greenland ice sheet retreats significantly can only be addressed through studies of past warm periods, such as the mid to late Cretaceous and the early Cenozoic, when the Arctic was ice-free and supported lush temperate rainforests and associated fauna. As with the other elements of a deep-time research agenda, improved dynamic models, more spatially and temporally resolved paleoclimate and paleontological datasets with high precision and chronological constraint, and data-model comparisons, are all critical components of future research efforts to better understand ecosystem processes and the dynamic interactions with changing climates.

STRATEGIES AND TOOLS TO IMPLEMENT THE RESEARCH AGENDA

The deep-time paleoclimate research agenda described above will require four key infrastructure and analytical elements, each of which is described in greater detail below:

1. Development and evaluation of new mineral and organic proxies and refinement of existing methods through calibration studies in modern systems and laboratories. Such efforts must be coupled with the development of multiproxy paleoclimate time series that are spatially resolved, of high temporal resolution, and of improved precision and accuracy.
2. Substantially increased investment in scientific continental drilling and continued support for scientific ocean drilling. Only recovery of high-quality cores can provide the requisite sample resolution and preservational quality to develop multiproxy archives for key paleoclimate targets across terrestrial-paralic-marine transects and latitudinal or longitudinal transects. The International Continental Drilling Program has a strong record of drilling a range of scientific objectives, including paleoclimate targets, but in contrast to strong support for this program by the European science community and other countries, U.S. support has

been at relatively low levels. Also, although U.S. leadership in scientific ocean drilling has been a major factor in the present understanding of past climates and climate-ocean linkages, recent funding cutbacks have jeopardized the potential for the oceanic component of the deep-time paleoclimate agenda described here to be realized.

3. Development of a new generation of models for paleoclimate studies, capable of focusing on past warm worlds and on extreme and/or abrupt climate events. Such new models will require unprecedented spatial resolution and additional capabilities to permit innovative data-model and model-model intercomparisons that are more consistent with Intergovernmental Program on Climate Change (IPCC) style assessments. This will maximize the potential for paleoclimate modeling studies to inform climate model development in general and for future climate simulations.

4. Substantially increased programmatic and financial support for the cultural and technological infrastructure that is needed for a "sea change" in the deep-time research culture—a shift away from single principal investigator (PI) or small collaborative projects to fully interdisciplinary synergistic research teams. Support for such research efforts will require a serious and committed investment in human and financial resources to establish large-scale, integrative programs for analyzing and archiving stratigraphic, sedimentological, geochemical, and paleontological datasets. A key ingredient will be the formation of deep-time "observatories" to unify researchers of disparate but complementary expertise to target specific processes or intervals of time, dedicated software engineering support and computational resources for model development and deep-time climate simulations, and professional development workshops and summer institutes for student training and early-career scientists.

5. An interdisciplinary collaboration "infrastructure" to foster broad-based collaborations of observation-based scientists and modelers for team-based studies of important paleoclimate time slices, incorporating climate and geochemical models; capabilities for the development, calibration, and testing of highly precise and accurate paleoclimate proxies; and the continued development of digital databases to store proxy data and facilitate multiproxy and record comparisons across all spatial and temporal scales. Making the transition from single researcher or small-group research efforts to a much broader-based interdisciplinary collaboration will be only possible through a modification of scientific research culture and will require substantially increased programmatic and financial support.

Improved Proxies and Multiproxy Records

One of the most important areas of paleoclimatology research is the need for improved constraints on past levels of radiative forcing and better estimates of long-term equilibrium climate sensitivity for previous warm periods and major climate transitions. Estimates of pCO_2 (through "paleobarometer" proxies) beyond the ice core records of the past 800,000 years, however, are inherently constrained by sizable uncertainties and the limits of sensitivity of marine or terrestrial proxies and/or of the numerical models of the long-term carbon cycle on which they are based. Furthermore, no proxies exist for greenhouse gases other than CO_2, such as methane. Similarly, the precision and accuracy of existing organic and mineral paleotemperature proxies are compromised by their calibrations solely to extant analogues and by incompletely understood biological and environmental controls on stable isotope and trace metal incorporation into mineral proxies and/or their sensitivity to postdepositional alteration. In addition, a broader ensemble of proxies for estimating past terrestrial surface and soil temperatures and seasonality of precipitation is much needed. Therefore, focused efforts to refine and develop proxies for these parameters are a critical element of an enhanced deep-time paleoclimatology initiative.

Improving existing and new proxies will require field and laboratory calibration studies in modern marine and terrestrial systems in order to increase their accuracy and further quantify and constrain uncertainties associated with estimates. Expansion of organic fossil-based CO_2 (e.g., plant stomatal indices, alkenones of marine haptophytes) and paleotemperature (biomarker) proxies to extinct taxa that dominate the deep-time record will require laboratory microcosm (growth chambers) studies that can evaluate biotic responses and geochemical feedbacks associated with changing greenhouse gas levels or air-water temperature. For mineral-based paleobarometers and continental paleotemperature proxies, calibration studies are needed in modern soil systems over a spectrum of landscapes and climate regimes in order (1) to better understand the influence of local climate, regional and soil hydrology, and soil productivity on soil CO_2 contents, temperature, and moisture—the input parameters for proxy transfer functions and pCO_2 calculations, and (2) to assess the sensitivity of proxy pCO_2 and temperature estimates to these soil parameters, including their seasonal variability. The Critical Zone Observatories initiative funded by the National Science Foundation (NSF) may offer opportunities to integrate such calibration studies within existing observatories.

Ultimately, comparison studies of plant and mineral proxy estimates that are characterized by differing sensitivities and uncertainties are required to test the accuracy, precision, and sensitivity of each of the proxies. In this broader context, however, several foci require continued and/or scaled-up research effort. First, studies of the taxonomic effects on mineral and organic biotic proxies are needed, as are collaborations between geochemists and paleobiologists for testing and applying biotic proxies because of the importance of recognizing and evaluating vital effects on proxy values. Second, continued development of biomarker proxies should be a high priority given their high precision and sensitivity and the fact that they appear to be diagenetically robust. Future efforts, however, need to include (1) critical evaluation of the potential to extend various biomarker approaches beyond the temporal range of the taxa for which they were developed (e.g., the alkenone method using marine haptophytes; Freeman and Pagani, 2005); and (2) more rigorous assessment of the sources, distribution, and preservation potential of various biomarkers through the deep-time geological record. Third, further evaluation of the effects of post-depositional alteration on mineral isotopic and geochemical compositions is needed, and this will require the use of emerging submicron imaging and analytical technology (e.g., scanning electron microscopy, nanoscale secondary ion mass spectrometry [Nano-SIMS], laser ablation inductively coupled plasma mass spectrometry coupled to the Australia National University's sample cell). Fourth, increased efforts for development of emerging paleotemperature proxies that are independent of biological effects and water composition are needed (e.g., Mg isotopes, clumped isotope thermometry). Overall, because of the multidisciplinary nature of calibration and assessment studies and the diversity of natural and man-made laboratories in which they would need to be carried out, these efforts to improve the precision, accuracy, and array of paleoclimate proxies will require broad-based collaboration and long-term monetary and human resource investment.

Ultimately, reconstructions of regional variation in climate parameters and ecosystem changes will require multiproxy, spatially highly resolved, and temporally calibrated datasets that can be compared across marine-paralic-terrestrial and latitudinal-longitudinal gradients as a function of time. Only such databases will notably advance science's understanding of the marine-terrestrial dynamics of carbon and water cycling in a warmer world and their role in regional hydroclimate and ecosystem variability. Such studies should be undertaken in "real-time" collaboration with deep-time climate modeling efforts using fully integrated terrestrial and marine climate parameters. In turn, collaborative observation-climate model studies are an essential mechanism for refining the interpretive utility of proxies. Continued development of interactive analytical databases that permit the integration of new proxy data about past climate parameters and boundary conditions, within an existing rock-based spatial and temporal framework, is critical to facilitate the integration and comparison of multiple proxy time series along latitudinal-longitudinal transects or time-specific regional-to-global reconstructions, as well as to provide the best quality and consistent boundary conditions for climate model sensitivity tests and climate simulations.

Deep-Time Drilling Transects

A transect-based deep-time drilling program designed to identify, prioritize, drill, and sample key paleoclimate targets—involving a substantially expanded continental drilling program and additional support for the existing scientific ocean drilling program—constitutes the second component of the recommended deep-time paleoclimate research strategy. Establishment of such a drilling program would fulfill two basic requirements for a successful deep-time paleoclimate initiative. Firstly, such a program would provide a venue for the U.S. scientific community to develop broader synergistic interactions, both national and international, through team-based cross-disciplinary research projects. This model has been particularly successful for scientific ocean drilling in its various iterations—the Deep Sea Drilling Project (DSDP), the Ocean Drilling Program (ODP), and currently the Integrated Ocean Drilling Program (IODP). This would represent a dramatic change in research approach for land-based researchers, as a large component of terrestrial deep-time paleoclimatological studies until now have been single-PI or small-group driven. This proposed deep-time paleoclimate drilling would substantially expand the scope of the existing International Continental Drilling Program and provide a complementary perspective to the oceanic focus of IODP, targeting a much broader and longer swath of Earth history, as well as providing an additional emphasis on the critical—but understudied—paralic zone that holds considerable potential for delineating marine-terrestrial linkages.

Secondly, the proposed deep-time continental drilling program would provide a platform from which to develop multiproxy records with the requisite spatial and temporal resolution and unprecedented continuity and preservation. This program would also offer a chronostratigraphic framework into which outcrop-based data could be integrated, thereby broadening the dimensions of the paleoclimatic record.

Although scientific ocean drilling has provided much of the basis for what is presently known about Neogene climate dynamics and ocean-climate linkages, there is still a pressing need for high-resolution sections that carry clear signals of orbital forcing in older parts of the record, particularly the Paleogene and Cretaceous. Sections representing the greenhouse intervals for climatically sensitive regions are still required, specifically in the Arctic and proximal to Antarctica. Only one core has been taken in deep time in the Arctic, and although recovery was relatively poor it has provided the lone constraint on sea surface temperature for the basin during the PETM (see Sluijs et al., 2006). Funding support for full IODP operations has recently declined, and many important scientific objectives—including deep-time paleoclimate proposals—have consequently not been able to be undertaken; additional funding sufficient for the type of full operational schedule that applied in the 1990s—when so much of the present knowledge of Neogene climates was obtained from ocean cores—is essential.

Facilities for deep-time continental drilling would ideally consist of two components—a smaller mobile drill for shallow (tens to hundreds of meters) drilling (see Box 5.1) and a dedicated deeper-drilling platform (see Box 5.2). Because of the costliness of deep drilling, it is essential that continental drilling projects broadly integrate scientific expertise and interests and create a research culture analogous to that which has developed for scientific ocean drilling.

The proposed expanded continental and ocean drilling would provide three fundamental elements of the scientific research agenda presented earlier in this chapter:

- **Temporal Continuity and Improved Temporal Resolution:** The continuity of drill cores—and the proxy time series developed from them—provide superb records, particularly when compared with outcrop-based studies where accessibility has traditionally been limited by erosion and burial, as well as by exposure-related mineralogical and chemical alteration. The continuity and preservational quality of drill cores would maximize the potential for deep-time climate proxy records that are chronostratigraphically resolved at the orbital to seasonal range throughout the geological record.
- **Spatial Resolution:** Drilling continental successions not only permits sampling where surface outcrops are lacking, but also permits paleoclimate reconstructions at the spatial resolution required by the scientific questions being asked and hypotheses being tested. This would substantially expand the latitudinal and geographic range over which proxy records can be developed and would permit direct comparison of the marine and terrestrial proxy records that record fundamentally different climate responses (e.g., localized orographic effects on continents versus distant and homogenized signals from ocean currents). The lack of such geographically linked proxy time series has perhaps been the most limiting factor in documenting critical intervals that record extreme and/or abrupt (nonlinear) climate changes. Furthermore, the resulting paleoclimate reconstructions would vastly improve the quality of data-model comparisons and sensitivity analysis of climate models.
- **Isochrony of Records:** The quality of observation-based paleoclimate reconstructions is ultimately dependent on establishing contem-

BOX 5.1
Orbital Forcing of Tropical Triassic Climate: the Newark Basin Coring Project

From 1990 to 1994, the Newark Basin Coring Project (Figure 5.1) recovered 6.7 km of continuous core of Late Triassic and Early Jurassic lacustrine and fluvial sedimentary deposits and intercalated lava flows from seven sites in the Newark Rift Basin in New Jersey (Kent et al., 1995; Olsen et al., 1996). The sedimentary cycles of climatic origin documented in these cores, when combined with magnetostratigraphic, biostratigraphic, and chemostratigraphic records, now provide a standard reference section for tropical continental climate during the Late Triassic and earliest Jurassic. This reference section has served as the basis for the astronomically tuned Late Triassic-Early Jurassic geomagnetic polarity timescale (Kent et al., 1995; Kent and Olsen, 1999), which in turn has been correlated to the standard marine Tethyan stages and substages (Channell et al., 2003; Muttoni et al., 2003; Ruhl et al., 2010). This unique framework permits the regional and global correlation essential to the understanding of deep-time global climate change and macroevolutionary patterns and the development of models of continental rifting. Notably, coring was the *only* way to recover a complete section of the basin because of the extremely discontinuous outcrop in the often urbanized and heavily vegetated region.

Key to the success of the project was an offset drilling strategy consisting of a transect of seven 1- to 1.5-km boreholes that took advantage of the tilted strata within the half-graben Newark Basin, with each borehole sampling a different part of the section and with about 25% stratigraphic overlap between stratigraphically adjacent boreholes. This method of producing a composite section significantly reduced costs compared with a single hole and greatly reduced the possibility of encountering faults that would have compromised the continuity of the record. It also avoided penetration into the ~300-m Palisade diabase sill that underlies most of the basin, taking advantage of the fact that the intrusion penetrates different stratigraphic levels in different areas.

The lacustrine portion of the record spans 25 million years and displays a full range of precession-related orbital cycles including ~20, ~100, and 405 thousand year cycles (Olsen and Kent, 1996, 1999) (Figure 5.2). In addition to supplying a metronome for temporal calibration of the geomagnetic polarity timescale derived from the cores, the length of the record allows quantitative calibration of cycles that differ in period from the present due to chaotic diffusion of the gravitational system of the solar system (Olsen and Kent, 1999). Paleomagnetic polarity correlation to other areas with datable ashes strongly corroborates the astrochronology (Furin et al., 2006; Olsen et al., 2010) and is allowing temporal correlations of tectonic and macroevolutionary events, such as the eruption of the CAMP and the end-Triassic extinction at the subprecessional level (Olsen et al., 2002, 2003; Deenen et al., 2010; Whiteside et al., 2010).

continued

BOX 5.1 Continued

FIGURE 5.1 Drilling in the Newark Basin, New Jersey, using an offset drilling method.
SOURCE: Photo and text courtesy of Paul Olsen, Lamont-Doherty Geological Observatory of Columbia University.

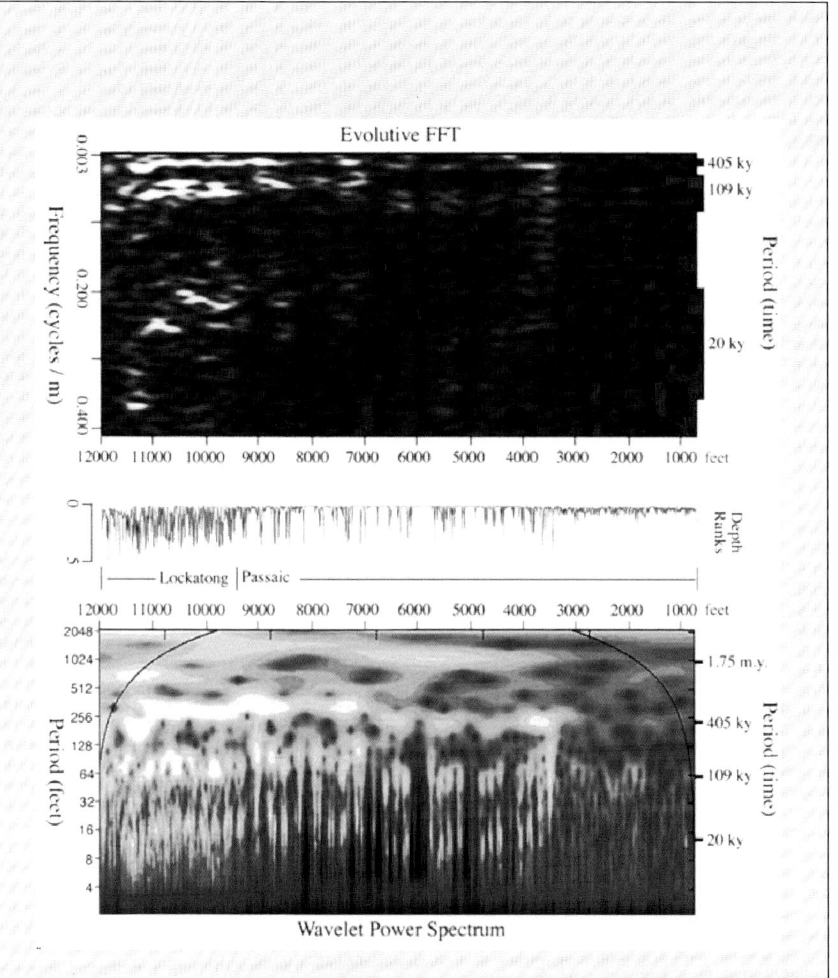

FIGURE 5.2 Color wavelet power spectrums for Neogene and Triassic lacustrine sediments of the Newark Basin showing a full range of precession-related (~20, ~100, 405 kyr) orbital cycles (Olsen and Kent, 1996, 1999).
SOURCE: Courtesy of Paul Olsen, Lamont-Doherty Geological Observatory of Columbia University.

BOX 5.2
Paleo-Rainforest Reconstruction Through Continental Drilling

The Denver Basin project (1997-2008) was an endeavor by the Denver Museum of Nature and Science to study the geology and paleontology of Late Cretaceous and Paleogene strata along the eastern flank of the Colorado Front Range. In the early 1990s, construction excavations in Denver and Colorado Springs uncovered diverse fossils that included dinosaurs, mammals, crocodiles, palm forests, and a fully tropical rainforest—apparently, the underpinnings of a major city contained the remains of a Paleogene greenhouse world. Despite the importance of the fossils and their richness, the rarity of natural surface outcrop and the extreme lateral variability of the strata precluded the construction of a useful stratigraphic framework and prevented a complete understanding of the paleoenvironment. To address this problem, the museum proposed to drill and core a 650-m drillhole in the center of the basin and to use magnetostratigraphy, radioisotope dating, palynostratigraphy, and the K-T boundary to construct a geochronological framework for basin interpretation. In 1999, with funding from NSF and the Colorado Water Conservation, the museum drilled the hole over a 7-week period (Figure 5.3), recovering more than 90 percent core from the 660-m well (Raynolds and Johnson, 2002). The cores permitted the precise dating of the fossil tropical rainforest, placing it at 1.4 Ma after the K-T boundary (Johnson and Ellis, 2002). The discovery of additional fossil rainforests revealed a mountain margin belt of high rainfall in which rainforests developed in the warm temperatures of the Paleocene.

FIGURE 5.3 Drilling rig collecting the 660-m Kiowa core from the Denver Basin project, which enabled the "Castle Rock Rainforest" to be dated at 64 Ma based on recovery of 35 ash layers spanning a 4-million-year section (Raynolds and Johnson, 2002).
SOURCE: Courtesy of Kirk Johnson, Denver Museum of Nature & Science.

poraneity in proxy records. Establishing isochrony of climate changes, in particular for past tipping points and abrupt climate change, is a key requirement for analysis of regional climate and ecosystem response to mean climate forcing, including global warming and the delineation of phasing between climate parameters. Additional drilling will greatly expand the potential for radiometric (e.g., EARTHTIME high-resolution istope dilution-thermal ionization mass spectrometry uranium-lead dating) and nonradiometric (e.g., orbital tuning) dating and correlation (e.g., magnetostratigraphy) of deep-time records by maximizing the preservation potential of volcanic ashes (see Box 5.2) and cyclic successions of extended duration that commonly are covered or deeply weathered in outcrop exposures.

To achieve these goals, the following elements will have to be part of the proposed expansion of deep-time continental drilling activities:

- A sustainable program of disciplinary planning workshops that would bring a range of different scientific communities together—including those that might not otherwise be engaged—to develop and plan drilling-based deep-time paleoclimatology projects.
- Creation and support of a database for archiving and sharing of data collected from drill cores and associated outcrop studies.
- Coordination of drilling efforts with existing U.S. and international drilling programs including the U.S. Geological Survey (USGS) and industry partners, with particular focus on predrilling site surveys that are costly but difficult to fund through traditional peer-reviewed proposal solicitations.

Improved Paleoclimate Modules and Models

The community's confidence in the ability of GCMs to forecast future regional and global climate changes depends in large part on the degree to which these models can simulate climates and feedbacks of deep-time climate systems. Modeling of ancient climates characterized by boundary conditions that may be substantially different from those of the present day (e.g., paleogeography, paleotopography, atmospheric pCO_2, solar luminosity parameters), however, presents a substantial challenge to the modeling community. An enhanced effort to model past warm intervals and periods of abrupt climate change across thresholds and possible tipping points is needed to produce models of future climate that can be adjusted to scenarios that include forcings or feedbacks not revealed by nearer-time paleoclimate reconstructions.

High Spatial Resolution

Although existing, more comprehensive Earth system models (ESMs or GCMs) offer many advantages, they are generally less capable at regional to local scales as a result of the coarseness of model resolution and the physical parameterizations, which are designed for larger geographic domains. This limitation is actually magnified for paleoclimate models, where following the lead of modern climate models can be hampered by lack of geographically extensive datasets and/or uncertain boundary conditions. In spite of these limitations, it is critical to run paleoclimate simulations at higher spatial resolutions both to capture high-resolution details required by sparser observational data and to maintain the ability for paleoclimate modeling conclusions to inform and evaluate future climate change simulations. Furthermore, as proxy datasets become increasingly more spatially resolved through additional drilling, an increase in model spatial resolution will be required to capture smaller-scale features. Downscaling techniques using either statistical approaches or nested fine-scale regional models will be required to better compare simulated climate variables to site-specific observational data.

Model Intercomparisons

Assessment of past climate changes also requires the regular comparison of results from a variety of independent climate models, as well as comparisons with proxy data or data-derived interpretations of climate change. The last four IPCC assessment reports have all incorporated multimodel ensemble forecasting techniques—in part, as a means of summarizing the large volume of results, but more importantly because different models, developed independently, do not produce identical results. The Fourth IPCC Assessment Report (IPCC, 2007) included a paleoclimate chapter for the first time, but there was little discussion of multimodel-data intercomparisons. Paleoclimate model intercomparison studies—which so far have been limited to the Last Glacial Maximum and mid-Holocene—do not produce identical results for any given past climate scenario. There is therefore a need for an expanded capability among modeling groups to compare simulations and create ensemble analyses in the same way that future climate change scenarios are examined. Although the first deep-time intercomparison project has recently begun—for the Pliocene—this is a small component of the vast array of possible model-model-data comparisons that are needed to better understand Earth's long-term climate sensitivity to CO_2 and anticipated regional climate changes.

Model-Data Tests

National and international model intercomparisons are particularly important for paleoclimate research because, unlike future climate experiments, there is an ability to evaluate model results using geological data. Deep-time records uniquely offer geological data, characterized by high climate signal-to-noise ratios and a broad spectrum of boundary conditions, to test how the various community models compare in their performance and sensitivities. Model-model-data comparisons, in particular for past warm climates characterized by elevated CO_2, provide a means of assessing not only the range of possible climate changes but also the credibility of climate model parameterizations in a way that future climate experiments are incapable of doing.

Recent advances in the capability of models to predict the spatial and temporal distribution of key environmental variables (temperature, precipitation and runoff, marine biological productivity, deep-water oxygen status) and, importantly, their proxies (e.g., water isotopic compositions, carbon isotopic compositions, organic carbon sedimentation rates) provide a more rigorous basis for the evaluation of hypotheses and understanding of drivers of environmental change in the geological past. New tools, however, are required to facilitate model-data comparison:

- Implementation of prognostic modules for proxies (e.g., isotopes) into comprehensive GCMs to facilitate direct comparisons between model and observations.
- Geographic information system (GIS)-based tools to align geological observations (geographically referenced to modern locations) with model simulations conducted with appropriate paleogeographies.
- Synthetic stratigraphic columns from the output of Earth system models run over millennia to millions of years, taking into consideration depositional, diagenetic, and erosional processes and thereby permitting direct comparison to actual stratigraphies. Ideally, these should also account for preservational biases.
- Refined dynamic vegetation models for climate modeling built on more comprehensive compilations of disparate paleobotanical data and improved knowledge of the composition and spatial distribution of vegetation on a global scale, prior to the evolution of angiosperms (Cretaceous) and grasslands (Cenozoic).
- Development of metadata techniques to ensure the utility and access of both model and observational electronically archived data by the larger scientific community involved in deep-time paleoclimatology studies.

Resource Requirements

Additional computational resources, however, will be needed because of the increased complexity, resolution, and length of integration of modeling runs. In addition to increased model spatial resolution, some paleoclimate simulations will have to be run for thousands of years to achieve near steady-state deep-ocean conditions and to assess climate variability at century timescales. This can be most effectively accomplished by the provision of dedicated computational resource for deep-time climate simulations. Such a resource could be created at one or more of the current national supercomputing centers that are already dedicated to the earth sciences (NSF, Department of Energy, National Aeronautics and Space Administration [NASA], or the National Oceanic and Atmospheric Administration [NOAA]), or at some new center. Three specific requirements for such a resource include:

- Dedicated technical support, in particular to incorporate chemical, biogeochemical, and ecological processes. Adaptation of Earth system models to paleoclimate applications has become progressively more challenging and requires sufficient technical support to enable a broader, multidisciplinary group of scientists to access these sophisticated models to address a diverse range of paleoclimate problems. Technical support is also required to reconfigure low-resolution climate models to run more efficiently on massively parallel computer architectures.
- Establishment of paleoclimate modeling archives that are better integrated with paleoclimate data archives, or better populating the growing IPCC-related archives (e.g., the Paleoclimate Modeling Intercomparison Project [PMIP]) with paleoclimate data repositories. This could include establishment of additional PMIP repository sites.
- An increased focus on the development and application of Earth system models of intermediate complexity (EMICs) that involve lower-resolution ocean models and either energy-moisture balance or coarse-resolution atmospheric models. Although the focus of EMIC application has been on the longer-term consequences of fossil fuel burning and the mechanisms of glacial-interglacial climate change (~10^4-year timescales), they have also been used to successfully simulate much longer events such as the PETM (Panchuk et al., 2008) and the end-Permian extinction (Meyer et al., 2008). Integration of intermediate-complexity models with three-dimensional comprehensive high-resolution Earth system modeling will ease the substantial demand on monetary and computational resources that the more complex three-dimensional models—which are essential for simulating the equilibrium states and short-term (decadal to century) variability in the climate system—currently require. Such an integrated approach will address the existing limitation of run times (less than several thousand

years) that are well below the duration of important climate perturbations in Earth history. To date, the U.S. contribution to the development of EMICs has been through collaboration with European and Canadian colleagues, and continued and expanded international collaboration—perhaps facilitated by the collaboration center proposed below—could yield an EMIC adapted to evaluate the mechanisms of environmental change in deep time (e.g., capable of simulating oceanic biogeochemical cycling under anoxic and euxinic conditions, using relevant paleogeographies). Future model development efforts could target the incorporation of subsystem models, such as EMICs within Earth system models.

Strategies for Fostering Focused Deep-Time Scientific Interaction

While the paleoclimate characteristics of past warm worlds and times of major climate transitions contained in the deep-time geological record constitute a substantially underdeveloped archive offering considerable potential for major scientific discoveries, such discoveries are unlikely to be made through single-PI disciplinary research or small-scale collaborative projects. For the full potential of the deep-time paleoclimate archive to be realized, it is critical to foster broad-based collaborations of observation-based scientists and climate modelers. Making the transition from single researcher or small-group research efforts to the broad-based interdisciplinary collaboration envisioned here will be possible only through a modification of the scientific research culture and will require substantially increased programmatic and financial support. The infrastructure needed to support scientific collaboration, cross-disciplinary syntheses, widespread and open data exchange, and cross-training of scientists and students will include, at a minimum, the following:

- The development of natural observatories—perhaps analogous to the NSF Critical Zone Observatories program—for team-based studies of important paleoclimate time slices or of landscapes that will permit the testing, calibration, and development of highly precise and accurate paleoclimate proxies (e.g., "Deep-Time" Critical Zone Observatory(ies)). Such deep-time observatories would serve to unify researchers of disparate but highly complementary expertise by targeting specific processes or intervals of time (e.g., the DETELON initiative by the paleobiology community). In order to develop the integrated sets of past-Earth boundary conditions critical to the success of GCMs—and currently a major limitation of climate modeling efforts—collaborative, cross-disciplinary teams would have to include software engineers and climate modelers as well as observational-based scientists with varying disciplinary expertise.
- Analytical support for interdisciplinary research through

BOX 5.3
Data Sharing in a Digital Age

The rock record serves as the primary long-term archive for many important physical, chemical, and biological processes, including the tempo and mode of organic evolution, the causes and consequences of global climate change, the rates and styles of crustal deformation and plate tectonics, and the origin and spatial and temporal distribution of mineral and energy resources. Although there exists a formidable body of knowledge on the distribution and character of rocks and the proxy data extracted from them, there is currently no framework for consolidating these data into a larger and interactive context or for analyzing them quantitatively across a range of time and spatial scales. Importantly, no such archive yet exists that can integrate with or accommodate paleoclimate modeling archives—a fundamental necessity for the proposed synergistic and interdisciplinary research approach to deep-time paleoclimatology.

Macrostratigraphy is a novel web-based data-sharing program (Figure 5.4) that uses gap-bound rock packages compiled separately at multiple geographic locations as a framework for integrating diverse geological and paleontological datasets and for analyzing quantitatively disparate data. Currently, this developing macrostratigraphic database consists minimally of the ages, thicknesses, lithologies, and nomenclatural hierarchies of 21,252 rock units from 821 geographic locations in North America, 1,168 rock units from 329 locations in New Zealand, and 7,124 lithologic packages from 132 locations in the deep sea. Macrostrat is fully integrated with the Paleobiology Database, thus serving as the scaffolding upon which to build a large-scale, integrative analytical framework for uniting stratigraphic, sedimentological, geochemical, and paleontological datasets spanning much of geoscience. Macrostrat has been utilized successfully to quantitatively analyze a wide range of geological questions, such as how the relative magnitudes of inorganic and organic carbon burial have fluctuated on a stage-to-stage basis throughout the Phanerozoic. The results of this example reveal the dominant influence of physically forced changes in sedimentation on carbon cycling on relatively short timescales, with implications for the relative cycling rates of terrestrial versus marine systems, for understanding the biological evolution of marine and terrestrial organisms, and for calibrating the link between carbon burial and global climate change. Ultimately, Macrostrat will provide a user-oriented web application that will enable participation of researchers widely throughout the community to facilitate data sharing and integration as well as continued development of new tools.

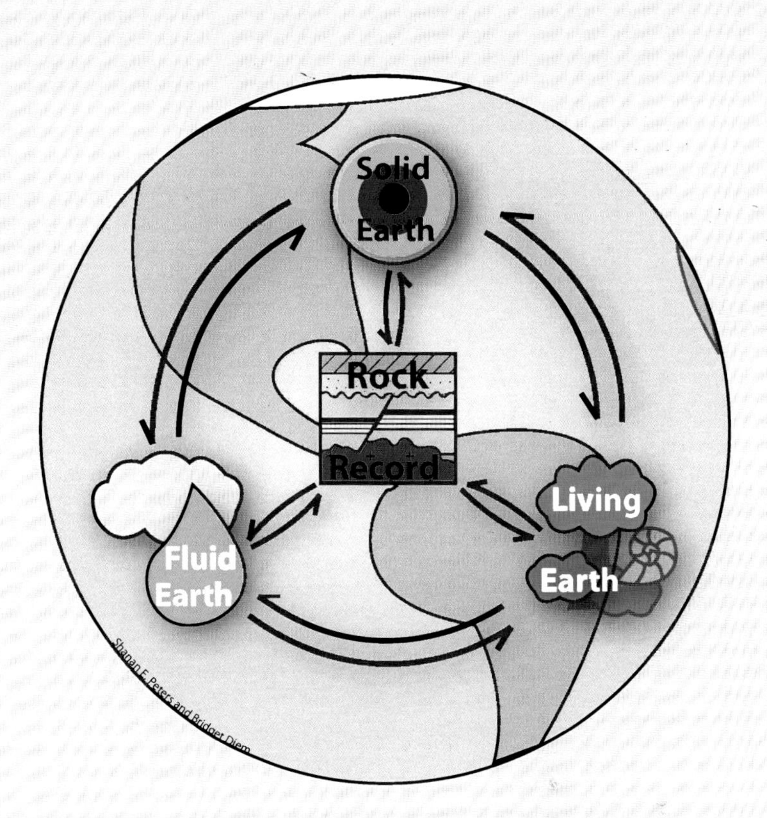

FIGURE 5.4 Logo for the macrostratigraphy database, Macrostrat. See http://macrostrat.geology.wisc.edu and http://strata.geology.wisc.edu/mibasin.
SOURCE: Courtesy of Shanan E. Peters, University of Wisconsin, Madison.

expanded efforts to develop new facilities (e.g., EARTHTIME geochronology laboratories) and enhanced linkages to existing structures (e.g., National Center for Nano-SIMS at the University of Wisconsin). Most importantly, it is critical that such facilities are made available to all interested scientific parties—an effort that will require proactive and strategic planning on the part of the funding agencies involved.

- Increased development efforts for large-scale, integrative analytical models for analyzing and archiving stratigraphic, sedimentological, geochemical, and paleontological datasets (see Box 5.3). Any such effort must incorporate plans to integrate with, or accommodate, paleoclimate model archives that can be fully integrated with geological, proxy, and paleontological data.
- To ensure that the collaborative opportunities offered are available to both researchers and "scientists in training," and to catalyze the cultural change in established and developing scientists, a structured mechanism for cross-disciplinary training of graduate students and early-career, and established scientists is necessary. Financial resources for professional development workshops and a summer institute(s) (perhaps on a rotating basis) in topics such as a modeling primer, overview and challenges of paleoclimate proxies, chronological techniques—all offered within the context of deep-time paradigms and unresolved problems—should be a high priority. Such institutes could easily be designed to incorporate secondary school teachers, museum specialists, and science journalists (see discussion in "Education and Outreach—Steps Toward a Broader Community Understanding of Climates in Deep Time" below).
- An emphasis on "virtual" collaborations would be cost-effective by removing the need for colocation of researchers, would be more in line with the comfort that younger researchers demonstrate with virtual interactions, and—by encouraging the interaction of widely distributed researchers—would help to emphasize that the issues being addressed are international in scope. Face-to-face meetings involving participants in a particular research endeavor would, of course, still be necessary on occasion, but these could perhaps take the form of annual workshops.

Most importantly, establishment of such a cultural and technological infrastructure will require acceptance and endorsement by both the scientific community and the funding agencies that support deep-time paleoclimatology and paleobiology-paleoecology studies. Without the addition of targeted new resources—in addition to existing programmatic resources—the scientific breakthroughs that can be made by this broad-based research community will be unlikely to come to fruition.

EDUCATION AND OUTREACH— STEPS TOWARD A BROADER COMMUNITY UNDERSTANDING OF CLIMATES IN DEEP TIME

Earth's deep-time climate history not only provides the context for scientists seeking to understand the Earth system, it also provides compelling opportunities for broad public outreach and education as it addresses details of Earth's natural long-term climate cycles. To capitalize on this opportunity, scientists and science communicators will have to overcome the challenge that understanding deep-time climate requires some appreciation of the subtleties of both climate science and geological time and an appreciation that Earth's preindustrial or even prehuman climate sets the baseline from which to evaluate the human contribution to climate change.

The public discussion regarding climate change and global warming is complex and fractious, in part reflecting the lack of adequate scientific literacy among the general public, an active campaign of antiscience disinformation, and insufficient efforts on the part of the scientific community to disseminate complex information in an effective manner. Similar things could be said about the public understanding of geological time, the age of the Earth, radioisotopic dating, and how scientists determine the age of events in Earth history. Despite these challenges, Earth's history is the source of useful and powerful metaphors and examples that have the potential to help people understand the significance of climate change in their time.

Challenges and Issues

The deep-time climate research community faces a number of challenges in bringing its insights to students, teachers, professors, scientific and media partners, policy makers, and the general public, and the following concepts and approaches are suggested to assist with education and outreach to convey the concepts and recommendations in this report. These are presented for each of the target audiences, with the challenges and issues associated with each audience and suggestions for audience-specific implementation.

Insights gleaned about Earth's climate system from the experiments of past climate extremes contained in the geological record both complement and expand those derived from climate studies of the more recent past. The study of deep-time paleoclimate integrates a large number of scientific disciplines because of the span of geological time, and as such it is not immediately intuitive to a nonspecialized audience. Although they are central to the practice and understanding of the science of deep-time climate, geological time, paleoclimate proxy analysis, and GCMs—with their attendant disciplinary jargon—have minimal traction with the public. Uncertainties in temporal resolution, patchwork spatial resolution, and incompletely calibrated climate proxies present challenges for conveying complex messages to the general public with sufficient simplification but without losing accuracy.

Finally, there are the difficulties inherent in any multidisciplinary field—communication between scientists in different fields is imperfect, leading to imperfect interpretations that can be propagated to conversations with policy makers and the public. Successful outreach and education need to be based on better integration of the scientific disciplines involved and improved transfer of data and knowledge between the different groups of scientists. Specifically, the observation and modeling communities need to improve their interdisciplinary communication, as well as broader communications with other scientists, to build a better pan-discipline understanding of what science knows about past climates—only then can these insights be effectively conveyed to broader audiences. Irrespective of these challenges, however, extinct animals and plants, ancient worlds, and natural disasters do resonate with people, and these elements all present good starting points for broader discussions about past climates

Audience-Specific Strategies and Examples

K-12

For elementary and secondary audiences, it is important to be aware of age-specific learning styles and state educational standards. Children are interested in dinosaurs and living animals, and these provide wonderful opportunities to discuss extinction and ecosystems. Extinct animals provide a pathway to discuss extinct landscapes and different biomes, and comparison of modern tropical, temperate, and polar biomes conveys the message that different animals live in places that have different weather.

Concepts of time are often challenging, and geological time is particularly difficult. This is amplified for younger children who are just beginning to understand the concept of time. The EARTHTIME Initiative[1] has recently created middle school curricula to address this issue, but the barriers to understanding are significant.

For young students, the excitement of exploring prehistoric worlds is more compelling and less scary than confronting the fear of global climate change. The recent *Ice Age* movies presented a climate change message in concert with charismatic ice age megafauna, introducing the topic in a fun, rather than threatening, manner. Children's love of dinosaurs is, in part, facilitated by the fact that they are scary but extinct. Climate change, while more conceptual than dinosaurs, is less threatening when studied

[1] See http://www.EARTHTIME.org.

as history than when presented as a looming threat. As always, creation of tools for teachers that adhere to state standards will result in more usable education assets.

For most states, geosciences are concentrated at the upper middle school level, and there is considerable potential to enhance existing curricula with deep-time climate science. The enhanced communication capabilities of the *Joides Resolution* drillship[2] have presented opportunities for live broadcasts to museums and classrooms. Active scientists should consider presenting simplified versions of their research and findings in classrooms, at teacher professional training sessions, and at national conventions such as the National Science Teachers Association meetings.

For high school audiences, awareness of global warming is high and rapidly increasing. Despite this, most curricula tend to be focused on the core sciences of biology, physics, and chemistry, with little opportunity to integrate the more synthetic earth and atmospheric sciences. Science fairs are arenas in which high school students can reach beyond the finite disciplines of their curricula, and students themselves can experience fieldwork through programs such as the Jason Project,[3] which places students in the field with the ability to broadcast back to their classrooms. In addition, a number of NSF-funded projects have generated web-based education tools. One such program that deals specifically with GCMs is the Educational Global Climate Modeling Program,[4] a collaboration between Columbia University and NASA's Goddard Institute for Space Studies, which allows web visitors to download and run simple climate models.

High school teachers can accrue very practical knowledge by participating in special training projects such as the IODP School of Rock Workshops,[5] which sends teachers to sea on the *Joides Resolution* drillship to learn about ocean science and seafloor coring. Museums offer teachers professional development on a variety of topics, such as the Denver Museum of Nature and Science Certification Program in Paleontology.[6] Inherent in all of these courses is the premise that high school teachers will be more effective if they have primary field experience.

Colleges and Universities

Colleges and universities provide a host of opportunities for students to understand deep-time climate topics through courses, fieldtrips, internships, visiting lectures and talks, and campus action groups. Earth and

[2] See http://joidesresolution.org/.
[3] See http://www.jason.org.
[4] See http://edgcm.columbia.edu.
[5] See http://www.iodp-usio.org/Education/SOR.html.
[6] See http://www.dmns.org.

planetary science departments can expand their ranks of potential majors by offering courses that combine traditional environmental science with paleontological and paleoclimatological content.

- Graduate students can expand their skills and knowledge by participating in interdisciplinary summer schools such as the Urbino Summer School of Paleoclimatology,[7] which bring together active scientists and diverse graduate students interested in paleoclimatology and modeling.
- Professional organizations such as the American Association of Petroleum Geologists and IODP support lecture tours by distinguished lecturers, and these can focus on paleoclimate themes.
- University and college faculty can improve their ability to communicate with the media and general public by specific training (e.g., the Aldo Leopold Leadership Program[8] at Stanford University).

General Public

The general public is barraged with global warming issues in the form of op-eds, letters to the editor, blogs, popular books, television shows, talk radio commentary, and newspaper and magazine ads from companies promoting green products to oil and gas companies discussing pathways to the future of energy. Despite this media blitz, very few people understand that the Earth's present climate is anomalously cold relative to the ice-free world of the Cretaceous to Eocene greenhouse. Nor could the typical citizen begin to articulate how tree rings, ice cores, and seafloor drilling relate to climate change. Efforts to explain "how scientists know what they know" are more likely to be received favorably than are proclamations about what will happen in the future.

The topic itself has become so polarized that some hosts consider global warming conversations akin to discussing religion or politics at the dinner table. To counter this, it makes sense to focus on relevant science rather than policy or practice. Efforts that intend to educate rather than advocate are more likely to be heard and understood by a diversity of audiences.

The deep-time observation and modeling communities both need to break into the popular science realm by emphasizing their more compelling and understandable elements. Great opportunities exist for the popularization of ice cap and ocean drilling, both of which occur in dramatic settings that are unfamiliar and interesting to the general public. These activities are great examples of science in action, and they show scientists

[7] See http://www.uniurb.it/ussp/.

[8] See http://www.leopoldleadership.org.

doing interesting activities in the pursuit of knowledge. Pathways to bring these activities to the public eye include television shows and series (e.g., Discovery Channel, National Geographic Channel, *The Daily Show, The Colbert Report*); enhanced presence on the radio (e.g., a paleoclimate feature on National Public Radio's (NPR's) Science Friday, Terry Gross interviews of proxy data analysts and modelers, Talk radio); Web and Web 2.0 tools (e.g., www.Ted.com; www.khanacademy.org/, www.story of stuff.com); Earth system and deep-time blogs using graduate students, scientists, and science writers (e.g., Andrew Revkin's *New York Times* Dot Earth blog[9]); popular books and magazine articles; the development of audience-tested museum exhibits; use of new media (e.g., Podcasts, Twitter, Facebook); and advertisement and amplification of credible climate websites (e.g., NOAA's climate website[10]).

Potential Collaborators

Science is now so specialized and complex that most scientists do not venture far from their particular research area. To obtain broader understanding of the potential offered by paleoclimate data and modeling within the larger climate discussion, it is important to create forums where scientists from different disciplines exchange information and perspectives. This is effectively done *within* disciplines by talks and symposia at national meetings (e.g., those hosted by the American Geophysical Union and the Geological Society of America) and *between* disciplines at meetings like those hosted by the American Association for the Advancement of Science. Opportunities to engage broader groups exist at venues such as industry conferences (e.g., the American Association of Petroleum Geologists) and environmental conferences (e.g., the Aspen Environment Forum) with the potential to build a broader collective understanding of the nature and reliability of proxy data and modeling.

Policy Makers

Ultimately, policy makers require scientifically credible and actionable data on which to base their policies. Faced with a diversity of opinions, they need credible sources of information. The IPCC made its findings very accessible by creating a simple, but multilingual, website that not only presented the report but also made its images and figures available for download as PowerPoint files. The creation of simple, but clear, collateral resources such as these should be a goal of deep-time research

[9] See http://dotearth.blogs.nytimes.com/.
[10] See http://globalchange.gov/; and http://www.realclimate.org.

projects. A good example is the Cenozoic climate curve of Zachos et al. (2001a), showing climate change over the last 70 million years using the proxy of ^{18}O in marine microfossils.

Specific Recommendations

One of the most significant, yet least understood, aspects of the deep-time climate record is the observation that Earth has moved between two major climate states—greenhouse and icehouse. Since the last transition between these two states occurred 34 million years ago at the end of the Eocene epoch, and the last time the Earth saw a transition from icehouse to greenhouse was nearly 300 million years ago, this is clearly a story told only by the deep-time record. This paleoclimate record contains facts that are startling to most people—there have been times when the poles were forested rather than being icebound; there were times when the polar seas were warm; there were times when tropical forests grew at midlatitudes; more of Earth history has been greenhouse than icehouse. Such relatively simple but relevant messages provide a straightforward mechanism for an improved understanding in the broader community of the importance of paleoclimate studies.

This message can be tailored to different audiences. For children, the simple comparison that dinosaurs lived in greenhouse conditions and mammoths lived in ice-house conditions can be an effective way to link a subject in which they are already interested to a phenomenon that should also interest them. With the first-order concept that the Earth's climate alternates between these two major climate states, it is then possible to find ways to discuss and explain shorter-wavelength variations in climate, such as the orbital parameters that drove the glacial and interglacial shifts of the Pleistocene or the oceanic changes that drive the El Niño cycles. From the perspective of deep time, it is possible to start with the big patterns and work toward the small ones, and this is exactly what does *not* happen when the story starts from the perspective of daily weather.

The deep-time record also includes examples of extreme climate events and transitions. These examples are very useful as tools to help explain the range of possibilities in the Earth's climate and to show how certain types of climate events can be abrupt, even when viewed from a human perspective. Examples such as the subdecadal warmings documented in the Greenland ice cores are useful to help people understand that just because something happened a long time ago, does not mean it took a long time to happen. With this realization, the deep-time record becomes a storehouse of useful and relevant examples. Ultimately, the goal of education and outreach from the deep-time perspective should be to help various audiences understand that the Earth has archived its climate history and that this archive—while not fully understood—is perhaps science's best tool to understand Earth's climate future.

Committing to Paleoclimate Education and Outreach

Given the poor state of the public's understanding of Earth sciences, and climate science in particular, it is time to commit to the ideal that education and outreach (E&O) cannot merely be afterthoughts to scientific research activities. By consigning E&O to a relatively minor role within science institutions and proposals, scientists have inadvertently but effectively cut off the public from understanding scientific research. The result has been that a significant percentage of the U.S. public distrusts or ignores scientific climate change information. Accordingly, rather than promote specific E&O programs, the committee recommends that there be a renewed commitment within every paleoclimate project to the dissemination and communication of results to students, teachers, and the public. The successful E&O activities associated with such programs as ANDRILL, IODP, and the Incorporated Research Institution for Seismology (IRIS) show that with an appropriately funded focal point for scientific interaction—a characteristic of each of these programs—it is possible to effectively convey rather complex scientific issues and scientific accomplishments to a broader audience. This reinforces the call for programmatic and funding support for broad-based interdisciplinary collaborations for deep-time paleoclimate science advanced in this report, because these collaborative focal points could easily include the type of dedicated E&O resources as the successful models noted above. Some existing E&O efforts for deep-time paleoclimatology have been summarized here, but these efforts have to be expanded and more such efforts should be established. In a field that suffers from chronically low resource allocations, education and outreach are suffering far more in the area of paleoclimate than in general climate education. However, students and the public have always had a particular affinity for Earth history and extreme events of the past, and accordingly this is a key area for attracting student and public attention to climate science in general.

6

Conclusions and Recommendations

The current rate of atmospheric increase of CO_2—~3 ppmv per year (IPCC, 2007)—is an order of magnitude or more greater than the increase in atmospheric CO_2 during the last deglaciation, when rapid retreat of northern hemisphere ice sheets led to rates of sea level rise of up to 5 m per century (Stanford et al., 2006). The last time the atmosphere contained CO_2 levels comparable to today's values, during the Pliocene, surface temperatures were on average ~3°C warmer, the Greenland ice sheet collapsed, and sea level rose by up to 30 m (Pagani et al., 2010; Seki et al., 2010). With the combination of continued burning of fossil fuels and the additional contribution of greenhouse gases to the atmosphere through positive feedbacks in the climate system, future atmospheric CO_2 levels could exceed 1,000 ppmv (Kump et al., 2009)—levels well above the stability threshold values for continental ice on Earth (Hansen et al., 2008). In fact, it is necessary to look back *at least* 34 million years—prior to the current icehouse—to examine climate change under such CO_2 levels. In this context, the magnitude and rate of the present greenhouse gas increase place the climate system in what could be one of the most severe increases in radiative forcing of the global climate system in Earth history.

To fully evaluate climate forcing feedbacks and tipping points that may characterize Earth's future, and to better understand climate change impacts and recovery, it is necessary to examine the records from past warm periods when there were similar magnitudes and rates of greenhouse gas forcing. The deep-time paleoclimatology record contains a rich archive of such warm worlds, and the associated transitions into and out of "greenhouse" conditions. For example, climate reconstructions of the end-Paleocene (~55 Ma), mid to late Cretaceous (~120 to 90 Ma), end-Triassic (~200 Ma), and Late Paleozoic (~300 to 251.2 Ma)—all periods associated with the massive release of greenhouse gases to the atmosphere—reveal dramatic changes in oceanic conditions and terrestrial climates. These changes brought about extensive restructuring of marine and terrestrial ecosystems that in many cases involved mass extinctions. These deep-time records also reveal that some of the feedbacks in the climate system may be unique to warmer worlds—and thus are not archived in more recent paleoclimate records—and accordingly might be expected to become increasingly relevant with continued warming. In particular, long-term feedbacks that are typically active on millennial scales are likely to become important at the human timescale, leading to substantial and abrupt (years to centuries) climate modifications. Reconstructions of past climates show that civilization has evolved in an anomalously stable period unrepresentative of the climate system's natural variability. Therefore, refining current understanding of climate dynamics (e.g., the range, rates, and magnitudes of feedbacks and change) during past periods of global warming, particularly times associated with epic deglaciations, is critical for assessing future risks. Improved understanding of climate dynamics will also aid efforts to mitigate the impact of continued warming on regional hydroclimates and water resources, ice sheet and sea level stability, and the health of marine and terrestrial ecosystems. Exciting research opportunities to help accomplish this task exist in the untapped potential of the deep-time geological record.

This report identifies a six-element research agenda designed to describe past climate variability and to better constrain how Earth's climate system has responded to episodes of changing greenhouse gas levels. The knowledge gained by this scientific agenda will be important for addressing questions regarding the projected rise in atmospheric CO_2 and the societal implications of this rise. The report also describes the research infrastructure necessary for successful implementation of the deep-time paleoclimatology agenda, as well as an education and outreach strategy designed to broaden our collective understanding of the unique perspective that the full range of the geological record provides for future climate change.

Improved Understanding of Climate Sensitivity and CO_2-Climate Coupling

Determining the sensitivity of Earth's mean surface temperature to increased greenhouse gas levels in the atmosphere is a key requirement for estimating the likely magnitude and effects of future climate change. The current understanding of climate sensitivity, defined on the basis of modern data and relatively recent paleoclimate records (≤20,000 years), is associated with large uncertainty (1.5 to ≥6°C). Positive feedbacks typically considered to have been active on longer timescales, but that may become increasingly relevant with continued warming, are not considered in these estimates. An improved definition of long-term equilibrium climate sensitivity—including more refined constraints on its lower boundary—over the full range and timescales of past radiative forcing is a major research priority. An associated focus is on gaining an improved understanding of how climate feedbacks and their role in amplifying climate change have varied with changes in greenhouse gas forcing. Accomplishing this objective will require the development of more accurate and precise paleo-CO_2 and paleotemperature proxies, as well as the development of new proxies for the full range of greenhouse gases. A complementary requirement is for high-resolution and high-precision time-series records, based on integrating multiproxy techniques. Data-model comparisons are needed to rigorously test non-CO_2 forcing mechanisms of global warming, as well as to refine the understanding of how the Earth's climate system would respond to increasing levels of atmospheric CO_2.

Climate Dynamics of Hot Tropics and Warm Poles

Recent climate modeling and deep-time paleoclimatology studies have demonstrated that the long-standing paradigm that the temperatures of tropical climates do not rise significantly during warm periods because of some type of temperature buffering mechanism is probably incorrect. Consequently, the mechanisms and feedbacks in the modern climate system that have controlled tropical and polar surface temperatures—ultimately leading to the existing relatively high pole-to-equator thermal gradient—may not operate in warmer worlds. A decreased latitudinal gradient in the future, which would almost certainly be associated with polar sea ice and continental ice sheet losses, would change atmospheric wind patterns and, in turn, ocean circulation—all having potential detrimental effects through teleconnections (Hay, 2010). To refine knowledge of the processes and climate feedbacks that may influence surface temperatures under higher atmospheric pCO_2, it is important that high-temporal-resolution, higher-precision proxy time series be developed across latitudinal transects, with a focus on reconstructing terrestrial-marine linkages. This will require a greatly increased effort in high-precision geochronological dating, coupled with substantially more spatially resolved proxy records. A more comprehensive understanding of the limits of tropical climate stability, the origin of anomalous polar warmth, and an understanding of how a weaker thermal gradient is established and maintained in warmer climate regimes will require further climate model development and deep-time data-model comparisons. These comparisons would also provide a much needed test of the efficacy of model projections of future climate.

Sea Level and Ice Sheet Stability in a Warm World

Study of the current icehouse climate state has provided better constraints on CO_2 and surface temperature threshold levels for ice sheet stability (Pagani et al., 2005, 2010; Pearson et al., 2009; Seki et al., 2010). Large gaps, however, remain in the understanding of ice sheet dynamics, with resulting limitations on the applicability of current coupled climate-ice sheet models. These issues highlight the uncertainties that still exist in projections regarding the timescales at which ice sheets would respond to continued warming and in understanding the influence of feedbacks not revealed by recent paleoclimate records or considered by future climate model projections (e.g., the projections used in IPPC, 2007). Consequently, the magnitude of sea level rise, once climate equilibrium is reached, remains elusive despite deep-time paleoclimate evidence that it could be substantially higher than model projections (Rohling et al., 2009). To markedly improve the understanding of climate–ice sheet–sea level dynamics relevant to a warming Earth, it will be necessary to probe deeper into Earth's history to the periods of truly catastrophic ice sheet collapse that accompanied past icehouse-to-greenhouse transitions. To fully exploit such deep-time archives will require radiometrically constrained and spatially resolved marine, paralic, and terrestrial records for both high and low latitudes. In addition, improved methods for deconvolving temperature and seawater $\delta^{18}O$ from proxy records are needed, as well as targeted efforts to couple land-ice component models with complex global climate models that are capable of integrating the atmospheric hydrological cycle.

Understanding the Hydrology of a Hot World

There is broad scientific consensus that one of the largest impacts of continued CO_2 forcing would be major regional climate modifications, with the likelihood of substantial societal impacts (e.g., water shortages, flooding). The insights gained from reconstructing the processes and climate feedbacks that influence surface temperatures under higher atmospheric pCO_2 levels are an important element of this research agenda, particularly because of the sensitivity of climate to small changes in high-latitude and tropical surface temperatures as a consequence of teleconnections. The deep-time geological record provides a critical and unique component of research focused on this issue, because it is the only source of information regarding how marine-terrestrial carbon and water cycle dynamics have influenced the global climate system during periods of radiative forcing comparable to those projected for the future, including periods of unipolar glacial or fully deglaciated greenhouse conditions. This will require a greatly expanded effort to develop linked marine-terrestrial records that are spatially resolved and of high temporal resolution, precision, and accuracy. New and improved quantitative estimates of paleoprecipitation, paleoseasonality, paleoaridity and humidity, and paleosoil conditions (including paleoproductivity) are critical components of this effort.

Understanding Tipping Points and Abrupt Transitions to a Warmer World

Studies of past climates and climate models show that Earth's climate system does not respond linearly to gradual CO_2 forcing, but rather responds by abrupt change as it is driven across climatic thresholds. Modern climate is changing very rapidly, and there is a possibility that Earth will soon pass thresholds that will lead to even more rapid changes in Earth's environments. Consequently, the question of how close Earth is to a tipping point, and when it could transition into a new climate state, is of critical importance. Because of their proven potential for capturing the dynamics of past abrupt changes, intervals of tipping-point climate transitions in the geological record—including past hyperthermals—should be the focus of future collaborative paleoclimate, paleoecological, and modeling studies. Such studies should lead to an improved understanding of how various components of the climate system responded to abrupt transitions, in particular during times when the rates of change were sufficiently large to imperil diversity. This research will also help determine whether there exist thresholds and feedbacks in the climate system of which we are currently unaware, especially in warm worlds and past icehouse-to-greenhouse transitions. Moreover, targeting such intervals for more detailed investigation is a critical requirement for constraining how long any abrupt climate change might persist.

Understanding Ecosystem Thresholds and Resilience in a Warming World

Both ecosystems and human society are highly sensitive to abrupt shifts in climate, because such shifts may exceed the tolerance of organisms and, consequently, have major effects on biotic diversity as well as human investments and societal stability. Modeling future biodiversity losses and biosphere-climate feedbacks, however, is inherently difficult because of the complex, nonlinear interactions with competing effects that result in an uncertain net response to climatic forcing. How rapidly biological and physical systems can adjust to abrupt climate change is a fundamental question accompanying present-day global warming. An important tool to address this question is to describe and understand the outcome of equivalent "natural experiments" in the deep-time geological record, particularly where the magnitude and/or rates of change in the global climate system were sufficiently large to threaten the viability and diversity of species, which at times led to mass extinctions. The paleontological record of the past few million years does not provide such an archive because it does not record catastrophic-scale climate and ecological events. As with the other elements of a deep-time research agenda, improved dynamic models, more spatially and temporally resolved datasets with high precision and chronological constraint, and data-model comparisons are all critical components of future research efforts to better understand ecosystem processes and dynamic interactions.

STRATEGIES AND TOOLS TO IMPLEMENT A DEEP-TIME CLIMATE RESEARCH AGENDA

Four key infrastructure and analytical elements will be required to implement this high-priority research agenda.

Improved Proxies and Multiproxy Records

Refinement of existing and development of new mineral and organic proxies for environmental and ecological parameters, coupled with an enhanced effort to chronologically calibrate targeted intervals with high-precision radiometric ages, are critical requirements for developing the spatially resolved, multiproxy paleoclimate and paleoecological time series described in the research agenda.

Despite exponential advances in the development of paleoclimate proxies over the past two decades, the precision and accuracy of existing organic and mineral paleotemperature and paleo-CO_2 proxies are compromised by their calibrations to extant analogues, by incompletely understood biological and environmental controls on geochemical signatures, and/or by their sensitivity to postdepositional alteration. Moreover, paleobarometer proxies are limited to CO_2, and there is a need for the existing very limited complement of proxies for estimating past terrestrial climatic conditions to be expanded and refined. A focused effort to improve existing proxies and develop new proxies is at the core of the proposed research agenda, in particular where the level of precision and accuracy—and thus the degree of uncertainty in inferred climate parameter estimates—can be quantified and significantly reduced. Such an effort will need to be highly collaborative, requiring calibration studies in modern marine and terrestrial environments as well as laboratory systems. The Critical Zone Observatories initiative funded by the National Science Foundation may offer opportunities to integrate such calibration studies into existing observatories. Ultimately, comparison studies of plant-mineral proxy estimates that are characterized by differing sensitivities and uncertainties are necessary to test the veracity and sensitivity of each of the proxies. Proxy development efforts must be complemented by studies that apply emerging imaging and analytical technology to critically evaluate the effects of postdepositional alteration on the compositions of isotopic and geochemical proxies.

Deep-Time Drilling Transects

The recovery of high-quality cores to provide the sample resolution and preservational quality needed to develop multiproxy archives for key paleoclimate targets across terrestrial-paralic-marine transects and latitudinal or longitudinal transects will require substantially increased investment in scientific continental drilling and continued support for scientific ocean drilling. Continental drilling will permit direct comparison of the marine and terrestrial proxy records that record fundamentally different climate responses (local and regional effects on continents compared with homogenized oceanic signals) and will provide the continuous records necessary for high-resolution dating of critical climate transition intervals.

The requirement for well-preserved and chronologically well-constrained proxy records with high spatial and temporal resolution and precision to analyze environmental and ecological systems in climate transition is a recurrent theme throughout the research agenda. A transect-based deep-time drilling program designed to identify, prioritize, drill, and sample key paleoclimate targets—involving a substantially expanded continental drilling program and additional support for the existing scientific ocean drilling program—is a high priority for implementing the recommended research agenda. Although scientific ocean drilling has provided much of the basis for what is presently known about Neogene climate dynamics and ocean-climate linkages, there is still a pressing need for high-resolution sections that carry clear signals of orbital forcing in older parts of the record, particularly the Paleogene and Cretaceous. Sections representing the greenhouse intervals for climatically sensitive regions are still required, specifically in the Arctic and proximal to Antarctica. Continental drilling of cyclic successions, of extended duration and with high potential for preservation of volcanic ashes, will greatly expand the opportunity for radiometric and nonradiometric dating and correlation, thereby facilitating comparison of paleoclimate records across marine-paralic-terrestrial gradients as a function of time.

Improved Paleoclimate Modules and Models

An enhanced paleoclimate modeling effort, with a focus on past warm worlds and extreme and/or abrupt climate events, is critical for refining scientific understanding of the complex dynamics of past climates and for producing models that can be adjusted to include forcings or feedbacks not revealed by shallower-time paleoclimate reconstructions.

As critical boundary conditions of the climate system—greenhouse gas concentrations, polar ice mass, distribution of biomes—change in the coming century, calibrations of climate models based on modern systems and the recent past will become increasingly less relevant. The deep-time geological record of past climates and major transitions provides the only test of climate models and their predictions against the range of background conditions most likely to be relevant to Earth's anticipated future climate state if emissions are not reduced. Modeling of ancient climates characterized by boundary conditions substantially different from those of the present day, however, presents a substantial challenge to the modeling community. In turn, how well such models simulate past climates and feedbacks inferred from deep time influences the community's confidence in the ability of global climate models to forecast future regional and global climate changes.

To that end, a markedly enhanced effort in deep-time paleoclimate modeling involving development of higher-resolution modules, improved parameterization of conditions relevant to future climate, and an emphasis on paleoclimate model intercomparisons and "next-generation" data-model comparisons is a fundamental component of the proposed research agenda. An increase in model spatial resolution will be required to capture smaller-scale features and regional climate changes comparable in scale to the spatially resolved geological data that can be obtained through continental drilling and proxy development. Deep-time data also uniquely offer the opportunity to carry out model-model-data comparisons for past warm climates characterized by elevated CO_2. Such comparisons will permit an assessment of the credibility of the performance and parameterizations of various community models in a way that future climate experiments are presently incapable of doing. Achieving this component of the deep-time initiative will require new tools to facilitate model-data comparisons (e.g., prognostic modules for proxies, geographic information system–based tools, refined dynamic vegetation models, metadata techniques), dedicated computational resources for deep-time climate simulations, and the development and application of Earth system models of intermediate complexity that can be integrated as subsystem models within more complex three-dimensional Earth system models.

Strategies for Fostering Focused Deep-Time Scientific Interaction

Implementing the research agenda described in this report will require a synergistic research culture among the broad range of disciplines that can contribute to solving the numerous puzzles of deep-time paleoclimatology, focusing on specific paleoclimate time slices as natural laboratories for team-based analyses of deep-time climates and their impact on Earth systems. Establishment of a cultural and technological infrastructure to support team-based projects offers the potential for discoveries unattainable by single-discipline research or even by more conventional integrated efforts.

Establishing the scientific collaboration, cross-disciplinary syntheses, widespread and open data exchange, cross-training of scientists and students, and dedicated and focused outreach activities required to address the research agenda described in this report will require the development of natural observatories for team-based studies of important paleoclimate time slices, incorporating climate and geochemical models; capabilities for the development, calibration, and testing of highly precise and accurate paleoclimate proxies; and the continued development of digital databases to store proxy data and facilitate multiproxy and record comparisons across all spatial and temporal scales. Such broad-based and interdisciplinary cultural and technological infrastructure will require acceptance and endorsement by both the scientific community and the funding agencies that support deep-time paleoclimatology and paleobiology-paleoecology studies. Without the addition of targeted new resources—in addition to existing programmatic resources—the scientific breakthroughs that can be made by this broad-based research community will be unlikely to ever come to fruition.

EDUCATION AND OUTREACH—STEPS TOWARD A BROADER COMMUNITY UNDERSTANDING OF CLIMATES IN DEEP TIME

Despite the potential and importance of the deep-time geological record, as articulated throughout this report, the public has minimal appreciation of the relevance of deep-time climates for Earth's future. This largely reflects the limited efforts by the scientific community to ensure that the importance and relevance of scientific efforts and results are conveyed to students, teachers, scientific and media partners, policy makers, and the general public. Barriers such as disciplinary jargon (geological time, paleoclimate proxies, and numeric climate models), imperfect interpretations and solutions created by uncertainties in temporal resolution, patchwork spatial resolution, and incompletely calibrated climate proxies, all present significant challenges for conveying complex messages to the general public with sufficient simplification but without losing accuracy. To resolve this issue, a strategy for education and outreach, to convey the lessons contained within deep-time records, should be tailored to the range of specific target audiences:

- K-12 elementary and secondary students. Museums are a key resource for educating students. Involving teachers in scientific endeavors can help demystify science and convey the excitement of scientific discovery, as well as being a method of disseminating scientific information.
- For colleges and universities, distinguished lecture tours, topical summer schools, and the integration of deep-time paleoclimatology into traditional and nontraditional earth science courses offer additional opportunities to convey the relevance of the deep-time record.
- To involve and educate the general public, the deep-time observation and modeling communities have opportunities to break into the popular science realm by emphasizing their more compelling and understandable elements. Immediate opportunities to illustrate "deep-time paleoclimatology in action" to the general public abound, whether the irreversible impact of past major climate changes on life, extreme glaciations and catastrophic deglaciations, or the mysteries of the ocean. The scientific community needs to proactively pursue pathways to the public provided by various multimedia opportunities.
- Potential scientific collaborators from the broader climate science community can obtain increased understanding of the potential offered by paleoclimate data and modeling through the creation or use of forums where scientists from different disciplines exchange information and perspectives. This can be effectively done *between* disciplines at meetings of broader groups (e.g., American Association for the Advancement of Science) and industry, environmental, ecology, and physical anthropology conferences.
- Policy makers require scientifically credible and actionable data on which to base their policies. Faced with a diversity of opinions, they need credible sources of information. This report and other National Research Council reports attempt to play this role, but in a much broader

sense the scientific community must strive to make the presentation of deep-time paleoclimate information as understandable as possible.

The paleoclimate record contains facts that are surprising to most people. For example, there have been times when the poles were forested rather than being icebound; there were times when the polar seas were warm; and there were times when tropical forests grew at midlatitudes. For the majority of Earth's history, the planet has been in a greenhouse state rather than in the current icehouse state. Such concepts provide an opportunity to help disparate audiences understand that the Earth has archived its climate history and that this archive, while not fully understood, provides crucial lessons to improving our understanding of Earth's climate future.

The possibility that our world is moving toward a "greenhouse" future continues to increase as anthropogenic carbon builds up in the atmosphere, providing a powerful motivation for understanding the dynamics of Earth's past "greenhouse" climates that are recorded in the deep-time geological record. It is the deep-time climate record that has revealed feedbacks in the climate system that are unique to warmer worlds—and thus are not archived in more recent paleoclimate records—and that might be expected to become increasingly relevant with continued warming. It is the deep-time record that has revealed the thresholds and tipping points in the climate system that have led to past abrupt climate change, including amplified warming, substantial changes in continental hydroclimate, catastrophic ice sheet collapse, and greatly accelerated sea level rise. Further, it is uniquely the deep-time record that has archived the full temporal range of climate change impacts on marine and terrestrial ecosystems, including ecological tipping points. An integrated research program—a deep-time climate research agenda—to provide a considerably improved understanding of the processes and characteristics over the full range of Earth's potential climate states offers great promise for informing individuals, communities, and public policy.

References

Abbot, D.S., and K.A. Emanuel, 2007. A tropical and subtropical land-sea-atmosphere drought oscillation mechanism. *Journal of the Atmospheric Sciences* 64: 4458-4466.

Abbot, D.S., and E. Tziperman, 2008. A high-latitude convective cloud feedback and equable climates. *Quarterly Journal of the Royal Meteorological Society* 134: 165-185.

ACIA (Arctic Climate Impact Assessment), 2004. Impacts of a Warming Arctic: Arctic Climate Impact Assessment. Cambridge, UK: Cambridge University Press. Available online at http://amap.no/acia/; accessed July 21, 2010.

Adams, D., M. Hurtgen, and B. Sageman, 2010. Volcanic activation of biogeochemical cascade regulates Oceanic Anoxic Event 2. *Nature Geoscience* 3: 201-204.

Adelson, J.M., G.R. Helz, and C.V. Miller, 2001. Reconstructing the rise of recent coastal anoxia; Molybdenum in Chesapeake Bay sediments. *Geochimica et Cosmochimica Acta* 65: 237-252.

Affek, H.P., M. Bar-Matthews, A. Ayalon, A. Matthews, and J.M. Eiler, 2008. Glacial/interglacial temperature variations in Soreq cave speleothems as recorded by "clumped isotope" thermometry. *Geochimica et Cosmochimica Acta* 72: 5351-5360.

Algeo, T.J., R.A. Berner, J.B. Maynard, and S.E. Scheckler, 1995. Late Devonian oceanic anoxic events and biotic crises: "Rooted" in the evolution of vascular land plants? *GSA Today* 5: 45, 64-66.

Algeo, T.J., S.E. Scheckler, and J.B. Maynard, 2001. Effects of early vascular land plants on weathering processes and global chemical fluxes during the Middle and Late Devonian. Pp. 213-236 in P. Gensel and D. Edwards (eds.), *Plants Invade the Land: Evolutionary and Environmental Perspectives*. New York: Columbia University Press.

Algeo, T., Y. Shen, T. Zhang, T. Lyons, S. Bates, H. Rowe, H., and T.K.T. Nguyen, 2008. Association of ^{34}S-depleted pyrite layers with negative carbonate δ^{13}C excursions at the Permian-Triassic boundary: Evidence for upwelling of sulfidic deep-ocean water masses. *Geochemistry Geophysics Geosystems* 9: Q04025.

Algeo, T.J., L. Hinnov, J. Moser, J.B. Maynard, E. Elswick, K. Kuwahara, and H. Sano, 2010. Changes in productivity and redox conditions in the Panthalassic Ocean during the latest Permian. *Geology* 38:187-190.

Allen, P.A., and J.L. Etienne, 2008. Sedimentary challenge to Snowball Earth. *Nature Geoscience* 1: 817-825.

Alley, N.F., and L.A. Frakes, 2003. First known Cretaceous glaciation: Livingston Tillite Member of the Cadna-owie Formation, South Australia. *Australian Journal of Earth Sciences* 50: 139-144.

Alley, R.B., J. Marotzke, W.D. Nordhaus, J.T. Overpeck, D.M. Peteet, R.A. Pielke, Jr., R.T. Pierrehumbert, P.B. Rhines, T.F. Stocker, L.D. Talley, and J.M. Wallace, 2003. Abrupt climate change. *Science* 299: 2005-2010.

Alley, R.B., P.U. Clark, P. Huybrechts, and I. Joughin, 2005. Ice sheet and sea level changes. *Science* 310: 456-460.

Alpert, P., D. Niyogi, R.A. Pielke, Sr., J.L. Eastman, Y.K. Xue, and S. Raman, 2006. Evidence for carbon dioxide and moisture interactions from the leaf cell up to global scales: Perspective on human-caused climate change. *Global and Planetary Change* 54: 202-208.

Anbar, A.D., and A.H. Knoll, 2002. Proterozoic ocean chemistry and evolution: A bioinorganic bridge? *Science* 297: 1137-1142.

Anderson, R.Y., 1982. A long geoclimatic record from the Permian. *Journal of Geophysical Research* 87: 7285-7294.

Archer, D., 2005. Fate of fossil fuel CO_2 in geologic time. *Journal of Geophysical Research* 110: C09S05, 6 pp.

Archer, D., 2009. *The Long Thaw: How Humans Are Changing the Next 100,000 Years of Earth's Climate*. Princeton, N.J.: Princeton University Press, 180 pp.

Archer, D., H. Kheshgi, and E. Maier-Riemer, 1997. Multiple timescales for neutralization of fossil fuel CO_2. *Geophysical Research Letters* 24: 405-408.

Archer, D., M. Eby, V. Brovkin, A. Ridgwell, L. Cao, U. Mikolajewicz, K. Caldeira, K. Matsumoto, G. Munhoven, A. Montenegro, and K. Tokos, 2009. Atmospheric lifetime of fossil fuel carbon dioxide. *Annual Review of Earth and Planetary Sciences* 37: 117-134.

Arnold, G.L., A.D. Anbar, J. Barling, and T.W. Lyons, 2004. Molybdenum isotope evidence for widespread anoxia in mid-Proterozoic oceans. *Science* 304: 87-90.

Arthur, M.A., H.C. Jenkyns, H.-J. Brumsack, and S.O. Schlanger, 1990. Stratigraphy, geochemistry, and paleoceanography of organic carbon-rich Cretaceous sequences. Pp. 75-119 in R.N. Ginsburg and B. Beaudoin (eds.), *Cretaceous Resources, Events and Rhythms: Background and Plans for Research*. Dordrecht, The Netherlands: Kluwer Academic.

Bains, S., R.M. Corfield, and R.D. Norris, 1999. Mechanisms of climate warming at the end of the Paleocene. *Science* 285: 724-727.

Baldocchi, D., F.M. Kelliher, T.A. Black, and P.G. Jarvis, 2000. Climate and vegetation controls on boreal zone energy exchange. *Global Change Biology* 6 (Suppl. 1): 69-83.

Bambach, R.K., 2006. Phanerozoic biodiversity mass extinctions. *Annual Review of Earth and Planetary Sciences* 34: 127-155.

Barclay, R.S., J.C. McElwain, and B.B. Sageman, 2010. Carbon sequestration activated by a volcanic CO_2 pulse during Ocean Anoxic Event 2. *Nature Geoscience* 3: 205-208.

Barker, P.F., G.M. Filippelli, F. Florindo, E.E. Martin, and H.D. Scher, 2007. Onset and role of the Antarctic Circumpolar Current. *Deep Sea Research Part II: Topical Studies in Oceanography* 54: 2388-2398.

Barrick, R.E., and M.J. Kohn, 2001. Comment: Multiple taxon–multiple locality approach to providing oxygen isotope evidence for warm-blooded theropod dinosaurs. *Geology* 29: 565-566.

Barron, E.J., 1987. Eocene equator-to-pole surface ocean temperatures: A significant climate problem? *Paleoceanography* 2: 729-739.

Barron, E.J., P.J. Fawcett, W.H. Peterson, D. Pollard, and S.L. Thompson, 1995. A "simulation" of mid-Cretaceous climate. *Paleoceanography* 10: 953-962.

Basinger, J.F., D.R. Greenwood, and T. Sweda, 1994. Early Tertiary vegetation of Arctic Canada and its relevance to paleoclimatic interpretation. Pp. 175-198 in M.C. Boulter and H.C. Fisher (eds.), *Cenozoic Plants and Climates of the Arctic.* NATO ASI Series. Berlin, Heidelberg: Springer-Verlag.

Beckmann, B., S. Flogel, P. Hofmann, M. Schulz, and T. Wagner, 2005. Orbital forcing of Cretaceous river discharge in tropical Africa and ocean response. *Nature* 437: 241-244.

Beerbower, R., J.A. Boy, W.A. DiMichele, R.A. Gastaldo, R. Hook, N. Hotton, III, T.L. Phillips, S.E. Scheckler, and W.A. Shear, 1992. Paleozoic terrestrial ecosystems. Pp. 205-325 in A.K. Behrensmeyer, J.D. Damuth, W.A. DiMichele, R. Potts, H.-D. Sues, and S.L. Wing (eds.), *Terrestrial Ecosystems Through Time: Evolutionary Paleoecology of Terrestrial Plants and Animals.* Chicago: University of Chicago Press.

Beerling, D.J., M. Harfoot, B. Lomax, and J.A. Pyle, 2007. The stability of the stratospheric ozone layer during the end-Permian eruption of the Siberian Traps. *Philosophical Transactions of the Royal Society A* 365: 1843-1866.

Beerling, D., R.A. Berner, F.T. Mackenzie, M.B. Harfoot, and J.A. Pyle, 2009. Methane and the CH_4 related greenhouse effect over the past 400 million years. *American Journal of Science* 309: 97-113.

Benson, L., S. Lund, R. Negrini, B. Linsley, and M. Zic, 2003. Response of North American Great Lakes to Dansgaard-Oeschger oscillations. *Quaternary Science Reviews* 22: 2239-2251.

Benton, M.J., and R.J. Twitchett, 2003. How to kill (almost) all life: The end-Permian extinction event. *Trends in Ecology and Evolution* 18: 358-365.

Berger, A., M.F. Loutre, and M. Crucifix, 2003. The Earth's climate in the next hundred thousand years (100 kyr). *Surveys in Geophysics* 24: 117-138.

Berner, R.A., 2004. *The Phanerozoic Carbon Cycle: CO_2 and O_2.* Oxford, UK: Oxford University Press. 160 pp.

Berner, R.A., 2006. GEOCARBSULF: A combined model for Phanerozoic atmospheric O_2 and CO_2. *Geochimica et Cosmochimica Acta* 70: 5653-5664.

Berner, R.A., 2009. Phanerozoic atmospheric oxygen: New results using the GEOCARBSULF model. *American Journal of Science* 309: 603-606.

Berner, R.A., and Z. Kothavala, 2001. GEOCARB III: A revised model of atmospheric CO_2 over Phanerozoic time. *American Journal of Science* 301: 182-204.

Bianchi, D., M. Zavatarelli, N. Pinardi, R. Capozzi, L. Capotondi, C. Corselli, and S. Masina, 2006. Simulations of ecosystem response during the sapropel S1 deposition event. *Palaeogeography, Palaeoclimatology, Palaeoecology* 235: 265-287.

Bice, K.L., D. Birgel, P.A. Meyers, K.A. Dahl, K.-U. Hinrichs, and R.D. Norris, 2006. A multiple proxy and model study of Cretaceous upper ocean temperatures and atmospheric CO_2 concentrations. *Paleoceanography* 21: PA2002, 17 pp.

Bidigare, R.R., A. Fluegge, K.H. Freeman, K.L. Hanson, J.M. Hayes, D. Hollander, J.P. Jasper, L. King, E.A. Laws, J. Milder, F.J. Millero, R.D. Pancost, B.N. Popp, P.A. Steinberg, and S.G. Wakeham, 1997. Consistent fractionation of ^{13}C in nature and in the laboratory: Growth rate effects in some haptophyte algae. *Global Biogeochemical Cycles* 11: 279-292.

Bijl, P.K., S. Schouten, A. Sluijs, G.-J. Reichart, J.C. Zachos, and H. Brinkhuis, 2009. Early Palaeogene temperature evolution of the southwest Pacific Ocean. *Nature* 461: 776-779.

Bishop, J.W., I.P. Montañez, E.L. Gulbranson, and P.L. Brenckle, 2009. The onset of mid-Carboniferous glacio-eustasy: Sedimentologic and diagenetic constraints, Arrow Canyon, Nevada. *Palaeogeography, Palaeoclimatology, Palaeoecology* 276: 217-243.

Blakey, R.C., 2008. Gondwana paleogeography from assembly to breakup: A 500 m.y. odyssey. Pp. 1-28 in C.R. Fielding, and J.L. Isbell (eds.), *Resolving the Late Paleozoic Ice Age in Time and Space.* Boulder, Colo.: Geological Society of America Special Paper 441.

Blisniuk, P.M., and L.A. Stern, 2005. Stable isotope paleoaltimetry: A critical review. *American Journal of Science* 305: 1033-1074.

Boer, G.J., K. Hamilton, and W. Zhu, 2005. Climate sensitivity and climate change under strong forcing. *Climate Dynamics* 24: 685-700.

Bornemann, A., R.D. Norris, O. Friedrich, B. Beckmann, S. Schouten, J.S. Sinninghe Damsté, J. Vogel, P. Hofmann, and T. Wagner, 2008. Isotopic evidence for glaciation during the Cretaceous supergreenhouse. *Science* 319: 189-192.

Botkin, D.B., H. Saxe, M.B. Araújo, R. Betts, R.H.W. Bradshaw, T. Cedhagen, P. Chesson, T.P. Dawson, J.R. Etterson, D.P. Faith, S. Ferrier, A. Guisan, A.S. Hansen, D.W. Hilbert, C. Loehle, C. Margules, M. New, M.J. Sobel, and D.R.B. Stockwell, 2007. Forecasting the effects of global warming on biodiversity. *Bioscience* 57: 227-236.

Bottjer, D.J., M.E. Clapham, M.L. Fraiser, and C.M. Powers, 2008. Understanding mechanisms for the end-Permian mass extinction and the protracted Early Triassic aftermath and recovery. *GSA Today* 18: 4-10.

Boucher, O., and U. Lohmann, 1995. The sulfate-CCN-cloud albedo effect: A sensitivity study with two general circulation models. *Tellus* 47B: 281-300.

Bowen, G.J., D.J. Beerling, P.L. Koch, J.C. Zachos, and T.A. Quattlebaum, 2004. A humid climate state during the Paleocene-Eocene Thermal Maximum. *Nature* 432: 495-499.

Breecker, D.O., Z.D Sharp, and L.D. McFadden, 2009. Seasonal bias in the formation and stable isotope composition of pedogenic carbonate in modern soils from central New Mexico, USA. *Geological Society of America Bulletin* 121: 630-640.

Brezinski, D.B., C.B. Cecil, V.W. Skema, and R. Stamm, 2008. Late Devonian glacial deposits from the eastern United States signal an end of the mid-Paleozoic warm period. *Palaeogeography, Paleoclimatology, Palaecology* 280: 143-151.

Brinkhuis, H., S. Schouten, M.E. Collinson, A. Sluijs, J.S. Sinninghe Damsté, G.R. Dickens, M. Huber, T.M. Cronin, J. Onodera, K. Takahashi, J.P. Bujak, R. Stein, J. van der Burgh, J.S. Eldrett, I.C. Harding, A.F. Lotter, F. Sangiorgi, H. van Konijnenburg-van Cittert, J.W. de Leeuw, J. Matthiessen, J. Backman, K. Moran, and the Expedition 302 Scientists, 2006. Episodic fresh surface waters in the Eocene Arctic Ocean. *Nature* 441: 606-609.

Broecker, W.S., 1999. What if the conveyor were to shut down? Reflections on a possible outcome of the great global experiment. *GSA Today* 9: 1-7.

Brusatte, S.L., M.J. Benton, M. Ruta, and G.T. Lloyd, 2008. The first 50 Myr of dinosaur evolution: Macroevolutionary pattern and morphological disparity. *Biology Letters* 4: 733-736.

Buggisch, W., M.M. Joachimski, G. Sevastopulo, and J.R. Morrow, 2008. Mississippian $\delta^{13}C_{carb}$ and conodont apatite $\delta^{18}O$ records—Their relation to the Late Palaeozoic glaciation. *Palaeogeography, Paleoclimatology, Palaecology* 268: 273-292.

Cadule, P., P. Friedlingstein, L. Bopp, S. Sitch, C.D. Jones, P. Ciais, S.L. Piao, and P. Peylin, 2010. Benchmarking coupled climate-carbon models against long-term atmospheric CO_2 measurements. *Global Biogeochemical Cycles* 24: GB2016, 24 pp.

Caldeira, K., and Wickett, M.E., 2003. Anthropogenic carbon and ocean pH. *Nature* 425: 365.

Caldeira, K., A. K. Jain, and M.I. Hoffert, 2003. Climate sensitivity uncertainty and the need for energy without CO_2 emission. *Science* 299: 2052-2054.

Came, R.E., J.M. Eiler, J. Veizer, K. Azmy, U. Brand, and C.R. Weidman, 2007. Coupling of surface temperatures and atmospheric CO_2 concentrations during the Palaeozoic era. *Nature* 449: 198-201.

Came, R. E., D. W. Oppo, W. B. Curry, and J. Lynch-Stieglitz, 2008. Deglacial variability in the surface return flow of the Atlantic meridional overturning circulation. *Paleoceanography* 23: PA1217.

Cane, M.A., 1998. A role for the tropical Pacific. *Science* 282: 59-60.

Cane, M.A., 2005. The evolution of El Niño, past and future. *Earth and Planetary Science Letters* 230: 227-240.

Cao, M., and Woodward, F.I., 1998. Dynamic responses of terrestrial ecosystem carbon cycling to global climate change. *Nature* 393: 249-252.

Capozzi, R., A. Negri, V. Picotti, E. Dinelli, S. Giunta, C. Morigi, P. Scotti, G. Lombi, and F. Marangoni, 2006. Mid-Pliocene warm climate and annual primary productivity peaks recorded in sapropel deposition. *Climate Research* 31: 137-144.

Carlson, A.E., A.N. LeGrande, D.W. Oppo, R.E. Came, G.A. Schmidt, F.S. Anslow, J.M. Licciardi, and E.A. Obbink, 2008. Rapid early Holocene deglaciation of the Laurentide ice sheet. *Nature Geoscience* 1: 620-624.

Cerling, T.E. 1991. Carbon dioxide in the atmosphere: Evidence from Cenozoic and Mesozoic paleosols. *American Journal of Science* 291: 377-400.

Cerling, T.E. 1992. Development of grasslands and savannas in East Africa during the Neogene. *Paleogeography, Paleoclimatology, Paleoecology* 97: 241-247.

Channell, J.E.T., H.W. Kozur, T. Sievers, R. Mock, R. Aubrecht, and M. Sykora, 2003. Carnian-Norian biomagnetostratigraphy at Silicka Brezova (Slovakia): Correlation to other Tethyan sections and to the Newark Basin. *Palaeogeography, Palaeoclimatology, Palaeoecology* 191: 65-109.

Chapin, F. S., III, M. Sturm, M.C. Serreze, J.P. McFadden, J.R. Key, A.H. Lloyd, A.D. McGuire, T.S. Rupp, A.H. Lynch, J.P. Schimel, J. Beringer, W.L. Chapman, H.E. Epstein, E.S. Euskirchen, L.D. Hinzman, G. Jia, C.-L. Ping, K.D. Tape, C.D.C. Thompson, D.A. Walker, and J.M. Welker, 2005. Role of land-surface changes in Arctic summer warming. *Science* 310: 657-660.

Chase, J.M., and M.A. Leibold, 2003. *Ecological Niches: Linking Classical and Contemporary Approaches*. Chicago: University of Chicago Press, 216 pp.

Chase, T.N., R.A. Pielke, T.G.F. Kittel, R.R. Nemani, and S.W. Running, 2000. Simulated impacts of historical land cover changes on global climate. *Climate Dynamics* 16: 93-105.

Cherchi, A., S. Masina, and A. Navarra, 2008. Impact of extreme CO_2 levels on tropical climate: A CGCM study. *Climate Dynamics* 31: 743-758.

Chikaraishi, Y., and H. Naraoka, 2003. Compound-specific δD-$\delta^{13}C$ analyses of *n*-alkanes extracted from terrestrial and aquatic plants. *Phytochemistry* 63: 361-371.

Clapham, M.E., and N.P. James, 2007. Climate-driven biotic change in the marine Permian of eastern Australia during the aftermath of the late Paleozoic Ice Age. *Geological Society of America Abstracts with Programs* 39: 354.

Clement, A.C., R. Seager, M.A. Cane, and S.E. Zebiak, 1996. An ocean dynamical thermostat. *Journal of Climate* 9: 2190-2196.

Clement, A.C., A.C. Baker, and J. Leloup, 2010. Climate change: Patterns of tropical warming. *Nature Geoscience* 3: 8-9.

Collins, M., 2005. El Niño- or La Niña-like climate change? *Climate Dynamics* 24: 89-104.

Cortese, G., R. Gersonde, C.-D. Hillenbrand, and G. Kuhn, 2004. Opal sedimentation shifts in the World Ocean over the last 15 Myr. *Earth and Planetary Science Letters* 224: 509-527.

Covey, C., 1991. Credit the oceans? *Nature* 352: 196-197.

Covey, C., and E. Barron, 1988. The role of ocean heat transport in climatic change. *Earth-Science Reviews* 24: 429-445.

Cox, P.M., R.A. Betts, C.D. Jones, S.A Spall, and I.J.Totterdell, 2000. Acceleration of global warming due to carbon-cycle feedbacks in a coupled climate model. *Nature* 408: 184-187.

Coxall, H.K. P.A. Wilson, H. Pälike, C.H. Lear, and J. Backman, 2005. Rapid stepwise onset of Antarctic glaciation and deeper calcite compensation in the Pacific Ocean. *Nature* 433: 53-57.

Cramer, B.S., and D.V. Kent, 2005. Bolide summer: The Paleocene/Eocene thermal maximum as a response to an extraterrestrial trigger. *Palaeogeography, Palaeoclimatology, Palaeoecology* 224: 144-166.

Crowley, T.J., and K.C. Burke (eds.), 1998. *Tectonic Boundary Conditions for Climate Reconstructions*. New York: Oxford University Press, 304 pp.

Crowley, T.J., and J.C. Zachos, 2000. Comparison of zonal temperature profiles for past warm time periods. Pp. 50-76 in B.T. Huber, K.G. Macleod, and S.L. Wing (eds.), *Warm Climate in Earth History*. Cambridge, UK: Cambridge University Press, 480 pp.

Cuffey, K.A., and S.J. Marshall, 2000. Substantial contributions to sea level rise during the last interglacial from the Greenland ice sheet. *Nature* 404: 591-594.

Das, S.B., I. Joughin, M.D. Behn, I.M. Howat, M.A. King, D. Lizarralde, and M.P. Bhatia, 2008. Fracture propagation to the base of the Greenland ice sheet during supraglacial lake drainage. *Science* 320: 778-781.

Davydov, V.I., J.L. Crowley, M.D. Schmitz, and V.I. Poletaev, 2010. High-precision U-Pb zircon age calibration of the global Carboniferous time scale and Milankovitch band cyclicity in the Donets Basin, eastern Ukraine. *Geochemistry Geophysics Geosystems* 11: Q0AA04, 22 pp.

Dawson, M.R., R.M. West, W. Langston, Jr., and J.H. Hutchison, 1976. Paleogene terrestrial vertebrates: Northernmost occurrence, Ellesmere Island, Canada. *Science* 192: 781-782.

De'ath, G., J.M. Lough, and K.E. Fabricius, 2009. Declining coral calcification on the Great Barrier Reef. *Science* 323: 116-119.

DeConto, R.M., and D. Pollard, 2003. Rapid Cenozoic glaciation of Antarctica induced by declining atmospheric CO_2. *Nature* 421: 245-249.

DeConto, R.M., W.W. Hay, S.L. Thompson, and J. Bergengren, 1999. Late Cretaceous climate and vegetation interactions: Cold continental interior paradox. Pp. 391-406 in E. Barrera and C. Johnson (eds.), *Evolution of the Cretaceous Ocean-Climate System*. Special Papers 332. Boulder, Colo.: Geological Society of America.

DeConto, R., D. Pollard, P.A. Wilson, H. Pälike, C.H. Lear, and M. Pagani, 2008. Thresholds for Cenozoic bipolar glaciation. *Nature* 455: 652-656.

Deenen, M.H.L., M. Ruhl, N.R. Bonis, W. Krijgsman, W.M. Kuerschner, M. Reitsma, and M.J. van Bergen, 2010. A new chronology for the end-Triassic mass extinction. *Earth and Planetary Science Letters* 291: 113-125.

Dekens, P.S., D.W. Lea, D.K. Pak, and H.J. Spero, 2002. Core top calibration of Mg/Ca in tropical foraminifera: Refining paleotemperature estimation. *Geochemistry Geophysics Geosystems* 3: 1022, 29 pp.

Dekens, P.S., A.C. Ravelo, M.D. McCarthy, and C.A Edwards, 2008. A 5 million year comparison of Mg/Ca and alkenone paleothermometers. *Geochemistry, Geophysics, Geosystems* 9: Q10001, 18 pp.

deMenocal, P.B., 1995. Plio-Pleistocene African climate. *Science* 270: 53-59.

Diaz, R.J., and R. Rosenberg, 2008. Spreading dead zones and consequences for marine ecosystems. *Science* 321: 926-929.

Dickens, G.R., J.R. O'Neil, D.K. Rea, and R.M. Owen, 1995. Dissociation of oceanic methane hydrate as a cause of the carbon isotope excursion at the end of the Paleocene. *Paleoceanography* 10: 965-971.

DiMichele, W.A., I.P. Montañez, C.J. Poulsen, and N.J. Tabor, 2009. Climate and vegetational regime shifts in the late Paleozoic ice age earth. *Geobiology* 7: 200-226.

Driese, S.G., L.C. Nordt, W. Lynn, C.A. Stiles, C.I. Mora, and L.P. Wilding, 2005. Distinguishing climate in the soil record using chemical trends in a vertisol climosequence form the Texas Coastal Prairie, and application to interpreting Paleozoic paleosols in the Appalachian Basin. *Journal of Sedimentary Research* 75: 340-353.

Duan, Y., S. Severmann, A.D. Anbar, T.W. Lyons, G.W. Gordon, and B.B. Sageman, 2010. Isotopic evidence for Fe cycling and repartitioning in ancient oxygen-deficient settings: Examples from black shales of the mid to late Devonian Appalachian basin. *Earth and Planetary Science Letters* 290: 244-253.

Dworkin, S.I., L. Nordt, and S. Atchley, 2005. Determining terrestrial paleotemperatures using the oxygen isotopic composition of pedogenic carbonate. *Earth and Planetary Science Letters* 237: 56-68.

Eagle, R.A., E.A. Schauble, A.K. Tripati, T. Tutken, R.C. Hulbert, and J.M. Eiler, 2010. Body temperatures of modern and extinct vertebrates from ^{13}C-^{18}O bond abundances in bioapatite. *Proceedings of the National Academy of Sciences USA* 107: 10377-10382.

Edmond, J.M., and Y. Huh, 2003. Non-steady state carbonate recycling and implications for the evolution of atmospheric P_{CO_2}. *Earth and Planetary Science Letters* 216: 125-139.

Eglinton, G., and R.J. Hamilton, 1967. Leaf epicuticular waxes. *Science* 156: 1322-1335.

Eglinton, T.I., and G. Eglinton, 2008. Molecular proxies for paleoclimatology. *Earth and Planetary Science Letters* 275: 1-16.

Ehlers, T.A., and C.J. Poulsen, 2009. Influence of Andean uplift on climate and paleoaltimetry estimates. *Earth and Planetury Science Letters* 281: 238-248.

Eiler, J.M., 2007. "Clumped-isotope" geochemistry—The study of naturally-occurring, multiply-substituted isotopologues. *Earth and Planetary Science Letters* 262: 309-327.

Ekart, D.D., T.E. Cerling, I.P. Montañez, and N.J. Tabor, 1999. A 400 million year carbon isotope record of pedogenic carbonate: Implications for paleoatmospheric carbon dioxide. *American Journal of Science* 299: 805-827.

Ekdahl, E.J., S.C. Fritz, P.A. Baker, C.A. Rigsby, and K. Coley, 2008. Holocene multidecadal- to millennial-scale hydrologic variability on the South American Altiplano. *Holocene* 18: 867-876.

Elderfield, H., and G. Ganssen, 2000. Past temperature and $\delta^{18}O$ of surface ocean waters inferred from foraminiferal Mg/Ca ratios. *Nature* 405: 442-445.

Elliott, W.P., and J.K. Angell, 1997. Variations of cloudiness, precipitable water, and relative humidity over the United States: 1973-1993. *Geophysical Research Letters* 24: 41-44.

Elrick, M., and L.A. Hinnov, 2007. Millennial-scale paleoclimate cycles recorded in widespread Palaeozoic deeper water rhythmites of North America. *Palaeogeography, Palaeoclimatology, Palaeoecology* 243: 348-372.

Emeis, K.-C., and H. Weissert, 2009. Tethyan-Mediterranean organic carbon-rich sediments from Mesozoic black shales to sapropels. *Sedimentology* 56: 247-266.

Erbacher, J., B.T. Huber, R.D. Norris, and M. Markey, 2001. Increased thermohaline stratification as a possible cause for an ocean anoxic event in the Cretaceous period. *Nature* 409: 325-327.

Eriksson, K.A., and E.L. Simpson, 2000. Quantifying the oldest tidal record: The 3.2 Ga Moodies Group, Barberton Greenstone Belt, South Africa. *Geology* 28: 831-834.

Erwin, D.H., 2006. Dates and rates: Temporal resolution in the deep time stratigraphic record. *Annual Review of Earth and Planetary Sciences* 34: 569-590.

Fairchild, I.A., and M.J. Kennedy, 2007. Neoproterozoic glaciation in the Earth system. *Journal of the Geological Society* 164: 895-921.

Fedorov, A.V., Dekens, P.S., McCarthy, M., Ravelo, A.C., deMenocal, P.B., Barreiro, M., Pacanowski, R.C., and Philander, S.G., 2006. The Pliocene paradox (mechanisms for a permanent El Niño). *Science* 312: 1485-1489.

Federov, A.V., C.M. Brierley, and K. Emanuel, 2010. Tropical cyclones and permanent El Niño in the Early Pliocene Epoch. *Nature* 463: 1066-1070.

Feldman, H.R., A.W. Archer, E.P. Kvale, C.R. Cunningham, C.G. Maples, and R.R. West, 1993. A tidal model of carboniferous Konservat-Lagerstaetten formation. *Palaios* 8: 485-498.

Ferguson, J.E., G.M. Henderson, M. Kucera, and R.E.M. Rickaby, 2008. Systematic change of foraminiferal Mg/Ca ratios across a strong salinity gradient. *Earth and Planetary Science Letters* 265: 153-166.

Fielding, C.R., T.D. Frank, L.P. Birgenheier, M.C. Rygel, A.T. Jones, and J. Roberts, 2008. Stratigraphic imprint of the late Palaeozoic Ice Age in eastern Australia: A record of alternating glacial and nonglacial climate regime. *Journal of the Geological Society* 165: 129-140.

Fine, M., and D. Tchenov, 2007. Scleractinian coral species survive and recover from decalcification. *Science* 315: 1811.

Fletcher, B.J., D.J. Beerling, S.J. Brentnall, and D.L. Royer, 2005. Fossil bryophytes as recorders of ancient CO_2 levels: Experimental evidence and a Cretaceous case study. *Global Biogeochemical Cycles* 19: GB3012, 13 pp.

Fletcher, B.J., S.J. Brentnall, C.W. Anderson, R.A. Berner, and D.J. Beerling, 2008. Atmospheric carbon dioxide linked with Mesozoic and early Cenozoic climate change. *Nature Geoscience* 1: 43-48.

Floegel, S., and T. Wagner, 2006. Insolation-control on the Late Cretaceous hydrological cycle and tropical African climate—Global climate modelling linked to marine climate records. *Palaeogeography, Palaeoclimatology, Palaeoecology* 235: 288-304.

Flower, B.P., and J.P. Kennett, 1995. Middle Miocene deepwater paleoceanography in the southwest Pacific: Relations with East Antarctic ice sheet development. *Paleoceanography* 10: 1095-1112.

Forest, C.E., 2007. Paleoaltimetry; A review of thermodynamic methods. *Reviews in Mineralogy and Geochemistry* 66: 173-193.

Foster, G.L., 2008. Seawater pH, pCO_2 and $[CO_3^{2-}]$ variations in the Caribbean Sea over the last 130 kyr: A boron isotope and B/Ca study of planktic foraminifera. *Earth and Planetary Science Letters* 271: 254-266.

Franks, P.J., and D.J. Beerling, 2009. Maximum leaf conductance driven by CO_2 effects on stomatal size and density over geologic time. *Proceedings of the National Academy of Sciences USA* 106: 10343-10347.

Freeman, K.H., and J.M. Hayes, 1992. Fractionation of carbon isotopes by phytoplankton and estimates of ancient CO_2 levels. *Global Biogeochemical Cycles* 6: 185-198.

Freeman, K.H., and M. Pagani, 2005. Alkenone-based estimates of past CO_2 levels: A consideration of their utility based on an analysis of uncertainties. Pp. 55-78 in J. Ehleringer, T. Cerling, and D. Dearing (eds.), *A History of Atmospheric CO_2 and Its Implications for Plants, Animals, and Ecosystems*. New York: Springer Science.

Fricke, H.C., and S.L. Wing, 2004. Oxygen isotope and paleobotanical estimates of temperature and $\delta^{18}O$-latitude gradients over North America during the early Eocene. *American Journal of Science* 304: 612-635.

Fricke, H.C., W.C. Clyde, J.R. O'Neil, and P.D. Gingerich, 1998. Evidence for rapid climate change in North America during the latest Paleocene Thermal Maximum: Oxygen isotope compositions of biogenic phosphate from the Bighorn Basin (Wyoming). *Earth and Planetary Science Letters* 160: 193-208.

Friedlingstein, P., P. Cox, R. Betts, L. Bopp, W. von Bloh, V. Brovkin, P. Cadule, S. Doney, M. Eby, I. Fung, G. Bala, J. John, C. Jones, F. Joos, T. Kato, M. Kawamiya, W. Knorr, K. Lindsay, H. D. Matthews, T. Raddatz, P. Rayner, C. Reick, E. Roeckner, K.-G. Schnitzler, R. Schnur, K. Strassmann, A. J. Weaver, C. Yoshikawa, and N. Zeng, 2006. Climate-carbon cycle feedback analysis: Results from the C^4MPI model intercomparison. *Journal of Climate* 19: 3337-3353.

Furin, S., N. Preto, M. Rigo, G. Roghi, P. Gianolla. J.L. Crowley, and S.A. Bowring, 2006. High-precision U-Pb zircon age from the Triassic of Italy: Implications for the Triassic time scale and the Carnian origin of calcareous nannoplankton and dinosaurs. *Geology* 34: 1009-1012.

Gale, A.S., J. Hardenbol, B. Hathway, W.J. Kennedy, J.R. Young, and V. Phansalkar, 2002. Global correlation of Cenomanian (Upper Cretaceous) sequences: Evidence for Milankovitch control on sea level. *Geology* 30: 291-294.

Gale, A.S., S. Voigt, B.B. Sageman, and W.J. Kennedy, 2008. Eustatic sea-level record for the Cenomanian (Late Cretaceous)—Extension to the Western Interior Basin, USA. *Geology* 35(11): 859-862.

Galeotti, S., G. Rusciadelli, M. Sprovieri, L. Lanci, A. Gaudio, and S. Pekar, 2009. Sea-level control on facies architecture in the Cenomanian-Coniacian Apulian margin (Western Tethys): A record of glacio-eustatic fluctuations during the Cretaceous greenhouse? *Palaeogeography, Palaeoclimatology, Palaeoecology* 276: 196-205.

Galeotti, S., S. Krishnan, M. Pagani, L. Lanci, A. Gaudio, J.C. Zachos, S. Monechi, G. Morelli, and L. Lourens, 2010. Orbital chronology of Early Eocene hyperthermals from the Contessa Road section, central Italy. *Earth and Planetary Science Letters* 290: 192-200.

Gallego-Torres, D., F. Martinez-Ruiz, A. Paytan, F.J. Jimenez-Espejo, and M. Ortega-Huertas, 2007. Pliocene-Holocene evolution of depositional conditions in the eastern Mediterranean: Role of anoxia vs. productivity at time of sapropel deposition. *Palaeogeography, Palaeoclimatology, Palaeoecology* 246: 424-439.

Gensel, P.G., and H.N. Andrews, 1987. The evolution of early land plants. *American Scientist* 75: 478-489.

Ghosh, P., C.N. Garzione, and J.M. Eiler, 2006. Rapid uplift of the altiplano revealed through ^{13}C-^{18}O bonds in paleosol carbonates. *Science* 311: 511-515.

Gillespie, A.R., S.C. Porter, and B.F. Atwater (eds.), 2004. *The Quaternary Period in the United States*. Amsterdam: Elsevier, 584 pp.

Gladwell, M., 2000. *The Tipping Point: How Little Things Can Make a Big Difference.* Boston: Little, Brown and Company, 279 pp.

Gooday, A.J., F. Jorissen, L.A. Levin, J.J. Middelburg, S.W.A. Naqvi, N.N. Rabalais, M. Scranton, and J. Zhang, 2009. Historical records of coastal eutrophication-induced hypoxia. *Biogeosciences* 6: 1707-1745.

Gordon, G.W., T.W. Lyons, G.L. Arnold, J. Roe, B.B. Sageman, and A.D. Anbar, 2009. When do black shales tell molybdenum isotope tales? *Geology* 37: 535-538.

Greenwood, D.R., and S.L. Wing, 1995. Eocene continental climates and latitudinal temperature gradients. *Geology* 23: 1044-1048.

Grice, K., C. Cao, G.D. Love, M.E. Böttcher, R.J. Twitchett, E. Grosjean, R.E. Summons, S.C. Turgeon, W. Dunning, and Y. Jin, 2005. Photic zone euxinia during the Permian-Triassic superanoxic event. *Science* 307: 706-709.

Grossman, E.L., T.E. Yancey, T.E. Jones, P. Bruckschen, B. Chuvashov, S.J. Mazzullo, and H.-S. Mii, 2008. Glaciation, aridification, and carbon sequestration in the Permo-Carboniferous: The isotopic record from low latitudes. *Palaeogeography, Palaeoclimatology, Palaeoecology* 268: 222-233.

Grotzinger, J.P., and A.H. Knoll, 1995. Anomalous carbonate precipitates: Is the Precambrian the key to the Permian? *Palaios* 10: 578-596.

Gussone, N., B. Hönisch, A. Heuser, A. Eisenhauer, M. Spindler, and C. Hemleben, 2009. A critical evaluation of calcium isotope ratios in tests of planktonic foraminifers. *Geochimica et Cosmochimica Acta* 73: 7241-7255.

Hansen, J.E., and M. Sato, 2001. Trends of measured climate forcing agents. *Proceedings of the National Academy of Sciences USA* 98: 14778-14783.

Hansen, J., M. Sato, P. Kharecha, D. Beerling, R. Berner, V. Masson-Delmotte, M. Pagani, M. Raymo, D.L. Royer, and J.C. Zachos, 2008. Target atmospheric CO_2: Where should humanity aim? *Open Atmospheric Science Journal* 2: 217-231.

Haq, B.U., and S.R. Schutter, 2008. A chronology of Paleozoic sea-level changes. *Science* 322: 64-68.

Hartmann, D.L., and M.L. Michelsen, 1993. Large-scale effects on the regulation of tropical sea surface temperature. *Journal of Climate* 6: 2049-2062.

Hassan, R., R. Scholes, and N. Ash, 2005. *Ecosystems and Human Well-Being: Current State and Trends*. Millennium Ecosystem Assessment Series Vol. 1. Washington, D.C.: Island Press, 948 pp.

Haug, G.H., D. Günther, L.C. Peterson, D.M. Sigman, K.A. Hughen, and B. Aeschlimann, 2003. Climate and the collapse of Maya civilization. *Science* 299: 1731-1735.

Haug, G.H., A. Ganopolski, D.M. Sigman, A. Rosell-Mele, G.E.A. Swann, R. Tiedemann, S.L. Jaccard, J. Bollmann, M.A. Maslin, M.J. Leng, and G. Eglinton, 2004. North Pacific seasonality and the glaciation of North America 2.7 million years ago. *Nature* 433: 821-825.

Hay, W.W., 2008. Evolving ideas about the Cretaceous climate and ocean circulation. *Cretaceous Research* 29: 725-753.

Hay, W.W., 2010. Can humans force a return to a "Cretaceous" climate? *Sedimentary Geology* 235: 5-26.

Hay, W.W., and DeConto, R.M., 1999. Comparison of modern and Late Cretaceous meridional energy transport and oceanology. Pp. 283-300 in E. Barrera and C.C. Johnson (eds.), *Evolution of the Cretaceous Ocean-Climate System*. Special Papers 332. Boulder, Colo.: Geological Society of America.

Hayes, J.M., H. Strauss, and A.J. Kaufman, 1999. The abundance of ^{13}C in marine organic matter and isotopic fractionation in the global biogeochemical cycle of carbon during the past 800 Ma. *Chemical Geology* 161: 103-125.

Haywood, A.M., M.A. Chandler, P.J. Valdes, U. Salzmann, D.J. Lunt, and H.J. Dowsett, 2009. Comparisons of mid-Pliocene climate predictions produced by the HadAM3 and GCMAM3 General Circulation Models. *Global and Planetary Change* 66: 208-224.

Head, J.J., J.I. Bloch, A.K. Hastings, J.R. Bourque, E.A. Cadena, F.A. Herrera, P.D. Polly, and C.A. Jaramillo, 2009. Giant boid snake from the Palaeocene neotropics reveals hotter past equatorial temperatures. *Nature* 457: 715-717.

Hegerl, G.C., T.J. Crowley, W.T. Hyde, and D.J. Frame, 2006. Climate sensitivity constrained by temperature reconstructions over the past seven centuries. *Nature* 440: 1029-1032.

Held, I.M., and B.J. Soden, 2006. Robust responses of the hydrological cycle to global warming. *Journal of Climate*, 19: 5686-5699.

Helliker, B.R., and S.L. Richter, 2008. Subtropical to boreal convergence of tree-leaf temperatures. *Nature* 454: 511-514.

Henderiks, J., and M. Pagani, 2007. Refining ancient carbon dioxide estimates: Significance of coccolithophore cell size for alkenone-based pCO_2 records. *Paleoceanography* 22: PA3202, 12 pp.

Hennessy, K.J., J.M. Gregory, and J.F.B. Mitchell, 1997. Changes in daily precipitation under enhanced greenhouse conditions. *Climate Dynamics* 13: 667-680.

Henriksson, A.S., M. Sarnthein, G. Eglinton, and J. Poynter, 2000. Dimethylsulfide production variations over the past 200 k.y. in the equatorial Atlantic: A first estimate. *Geology* 28: 499-502.

Hermoso, M., F. Minoletti, L. Le Callonnec, H.C. Jenkyns, S.P. Hesselbo, R.E.M. Rickaby, M. Renard, M. de Rafélis, and L. Emmanuel, 2009. Global and local forcing of early Toarcian seawater chemistry: A comparative study of different paleoceanographic settings (Paris and Lusitanian basins). *Paleoceanography* 24: PA4208.

Hesselbo, S.P., D.R. Gröcke, H.C. Jenkyns, C.J. Bjerrum, P. Farrimond, H.S. Morgans Bell, O.R. Green, 2000. Massive dissociation of gas hydrate during a Jurassic oceanic anoxic event. *Nature* 406: 392-395.

Hesselbo, S.P., H.C. Jenkyns, L.V. Duarte, and L.C.V. Oliveira, 2007. Carbon-isotope record of the Early Jurassic (Toarcian) Oceanic Anoxic Event from fossil wood and marine carbonate (Lusitanian basin, Portugal). *Earth and Planetary Science Letters* 253: 455-470.

Hickey, L.J., R.M. West, M.R. Dawson, and D.K. Choi, 1983. Arctic terrestrial biota: Paleomagnetic evidence of age disparity with mid-northern latitudes during the Late Cretaceous and Early Tertiary. *Science* 221: 1153-1156.

Higgins, J.A., and D.P. Schrag, 2006. Beyond methane: Towards a theory for the Paleocene-Eocene Thermal Maximum. *Earth and Planetary Science Letters* 245: 523-537.

Hinnov, L.A., and J.G. Ogg, 2007. Cyclostratigraphy and the astronomical time scale. *Stratigraphy* 4: 239-251.

Hochuli, P.A., A.P. Menegatti, H. Weissert, A. Riva, E. Erba, and I. Premoli Silva, 1999. Episodes of high productivity and cooling in the early Aptian Alpine Tethys. *Geology* 27: 657-660.

Hodell, D.A., M. Brenner, and J.H. Curtis, 2005. Terminal Classic drought in the northern Maya lowlands inferred from multiple sediment cores in Lake Chichancanab (Mexico). *Quaternary Science Reviews* 24: 1413-1427.

Hoegh-Guldberg, O., 1999. Climate change, coral bleaching and the future of the world's coral reefs. *Marine and Freshwater Research* 50: 839-866.

Hoegh-Guldberg, O., P.J. Mumby, A.J. Hooten, R.S. Steneck, P. Greenfield, E. Gomez, C.D. Harvell, P.F. Sale, A.J. Edwards, K. Caldeira, N. Knowlton, C.M. Eakin, R. Iglesias-Prieto, N. Muthiga, R.H. Bradbury, A. Dubi, and M.E. Hatziolos, 2007. Coral reefs under rapid climate change and ocean acidification. *Science* 14: 1737-1742.

Hoffman, P.F., and D.P. Schrag, 2002. The snowball Earth hypothesis: Testing the limits of global change. *Terra Nova* 14: 129-155.

Hoffman, P.F., J.A. Kaufman, and G.P. Halverson, 1998. Comings and goings of global glaciations on a Neoproterozoic carbonate platform in Namibia. *GSA Today* 8: 1-9.

Holbourn, A., W. Kuhnt, M. Schulz, J.-A. Flores, and N. Anderson, 2007. Orbitally-paced climate evolution during the middle Miocene Monterey carbon-isotope excursion. *Earth and Planetary Science Letters* 261: 534-550.

Holmden, C., R.A. Creaser, K. Muehlenbachs, S.A. Leslie, and S.M. Bergström, 1998. Isotopic evidence for geochemical decoupling between ancient epeiric seas and bordering oceans: Implications for secular curves. *Geology* 26: 567-570.

Hoogakker, B.A.A., G.P. Klinkhammer, H. Elderfield, E.J. Rohling, and C. Hayward, 2009. Mg/Ca paleothermometry in high salinity environments. *Earth and Planetary Science Letters* 284: 583-589.

Horton, D.E., and C.J. Poulsen, 2009. Paradox of late Paleozoic glacioeustasy. *Geology* 37: 715-718.

Horton, D.E., C.J. Poulsen, and D. Pollard, 2007. Orbital and CO_2 forcing of late Paleozoic continental ice sheets. *Geophysical Research Letters* 34: L19708.

Horton, D.E., C.J. Poulsen, and D. Pollard, 2010. Influence of high-latitude vegetation feedbacks on late Palaeozoic glacial cycles. *Nature Geoscience* 3: 572-577.

Hotinski, R.M., K.L. Bice, L.R. Kump, R.G. Najjar, and M.A. Arthur, 2001. Ocean stagnation and end-Permian anoxia. *Geology* 29: 7-10.

Huber, B.T., 2002. Deep-sea paleotemperature record of extreme warmth during the Cretaceous. *Geology* 30: 123-126.

Huber, M., 2008. A hotter greenhouse? *Science* 321: 353-354.

Huber, M., 2009. Climate change: Snakes tell a torrid tale. *Nature* 457: 669-671.

Huber, M., and R. Caballero, 2003. Eocene El Niño: Evidence for robust tropical dynamics in the "hothouse." *Science* 299: 877-881.

Huber, B.T., D.A. Hodell, and C.P. Hamilton, 1995. Middle-late Cretaceous climate of the southern high latitudes—Stable isotopic evidence for minimal equator-to-pole thermal gradients. *Geological Society of America Bulletin* 107: 1164-1191.

Immenhauser, A., T.F. Nägler, T. Steuber, and D. Hippler, 2005. A critical assessment of mollusk $^{18}O/^{16}O$, Mg/Ca, and $^{44}Ca/^{40}Ca$ ratios as proxies of Cretaceous seawater temperature seasonality. *Palaeogeography, Palaeoclimatology, Palaeoecology* 215: 221-237.

IPCC (Intergovernmental Panel on Climate Change), 1995. *Climate Change 1995: IPCC Second Assessment*. Cambridge, UK: Cambridge University Press.

IPCC (Intergovernmental Panel on Climate Change), 2001. *Climate Change 1995: IPCC Third Assessment*. Cambridge, UK: Cambridge University Press.

IPCC (Intergovernmental Panel on Climate Change), 2007. Climate Change 2007: The Physical Science Basis; Summary for Policymakers: Contribution of Working Group 1 to the Fourth Assessment Report of the Intergovernmental Panel on Climate Change, S. Solomon, D. Qin, M. Manning, Z. Chen, M. Marquis, K.B. Averyt, M. Tignor, and H.L. Miller (eds.). Cambridge, UK: Cambridge University Press, 996 pp.

Isozaki, Y., 1997. Permo-Triassic boundary superanoxia and stratified superocean: Records from lost deep sea. *Science* 276: 235-238.

Ivany, L.C, B.H. Wilkinson, K.C. Lohmann, E.R. Johnson, B.J. McElroy, and G.J. Cohen, 2004. Intra-annual isotopic variation in *Venericardia* bivalves: Implications for early Eocene temperature, seasonality, and salinity on the U.S. Gulf Coast. *Journal of Sedimentary Research* 74: 7-19.

Jackson, J.B.C., M.X. Kirby, W.H. Berger, K.A. Bjorndal, L.W. Botsford, B.J. Bourque, R.H. Bradbury, R. Cooke, J. Erlandson, J.A. Estes, T.P. Hughes, S. Kidwell, C.B. Lange, H.S. Henihan, J.M. Pandolfi, C.H. Peterson, R.S. Steneck, M.J. Tegner, and R.R. Warner, 2001. Historical overfishing and the recent collapse of coastal ecosystems. *Science* 293: 629-638.

Jahren, A.H., and L.S.L. Sternberg, 2002. Eocene meridional weather patterns reflected in the oxygen isotopes of Arctic fossil wood. *GSA Today* 12: 4-9.

Jahren, A.H., and L.S.L. Sternberg, 2003. Humidity estimate for the middle Eocene Arctic rain forest. *Geology* 31: 463-466.

Jahren, A.H., and L.S.L. Sternberg, 2008. Annual patterns within tree rings of the Arctic middle Eocene (ca. 45 Ma): Isotopic signatures of precipitation, relative humidity, and deciduousness. *Geology* 36: 99-102.

Jahren, A.H., N.C. Arens, G. Sarmiento, J. Guerrero, and R. Amundson, 2001. Terrestrial record of methane hydrate dissociation in the Early Cretaceous. *Geology* 29: 159-162.

Jahren, A.H., M.C. Byrne, H.V. Graham, L.S.L. Sternberg, and R.E. Summons, 2009. The environmental water of the middle Eocene Arctic: Evidence from δD, $\delta^{18}O$ and $\delta^{13}C$ within specific compounds. *Palaeogeography, Palaeoclimatology, Palaeoecology* 271: 96-103.

Jenkyns, H.C., 1988. The Early Toarcian (Jurassic) anoxic event: Stratigraphic, sedimentary, and geochemical evidence. *American Journal of Science* 288: 101-151.

Jenkyns, H.C., A. Matthews, H. Tsikos, and Y. Erel, 2007. Nitrate reduction, sulfate reduction, and sedimentary iron isotope evolution during the Cenomanian-Turonian oceanic anoxic event. *Paleoceanography* 22: PA3208, 17 pp.

Joachimski, M.M., P.H. von Bitter, and W. Buggisch, 2006. Constraints on Pennsylvanian glacioeustatic sea-level changes using oxygen isotopes of conodont apatite. *Geology* 34: 277-280.

Johnson, K.R., and B. Ellis, 2002. A tropical rainforest in Colorado 1.4 million years after the Cretaceous-Tertiary boundary. *Science* 296: 2379-2383.

Joughin, I., S.B. Das, M.A. King, B.E. Smith, I.M. Howat, and T. Moon, 2008. Seasonal speedup on the western flank of the Greenland ice sheet. *Science* 320: 781-784.

Jovane, L., M. Sprovieri, F. Florindo, G. Acton, R. Coccioni, B. Dall'Antonia, and J. Dinarès-Turell, 2007. Eocene-Oligocene paleoceanographic changes in the stratotype section, Massignano, Italy: Clues from rock magnetism and stable isotopes. *Journal of Geophysical Research* 112: B11101, 16 pp.

Kah, L.C., T.W. Lyons, and T.D. Frank, 2004. Low marine sulphate and protracted oxygenation of the Proterozoic biosphere. *Nature* 431: 834-838.

Katz, M.E., K.G. Miller, J.D. Wright, B.S. Wade, J.V. Browning, B.S. Cramer, and Y. Rosenthal, 2008. Stepwise transition from the Eocene greenhouse to the Oligocene icehouse. *Nature Geoscience* 1: 329-334.

Kearsey, T., R.J. Twitchett, G.D. Price, and S.T. Grimes, 2008. Isotope excursions and palaeotemperature estimates from the Permian/Triassic Boundary in the Southern Alps (Italy). *Paleogeography, Paleclimatology, Paleoecology* 279: 29-40.

Keeling, R.F., A. Körtzinger, and N. Gruber, 2010. Ocean deoxygenation in a warming world. *Annual Review of Marine Science* 2: 199-229.

Kendall, B., R.A. Creaser, C.R. Calver, T.D. Raub, and D.A.D. Evans, 2009. Correlation of Sturtian diamictite successions in southern Australia and northwestern Tasmania by Re-Os black shale geochronology and the ambiguity of "Sturtian"-type diamictite-cap carbonate pairs as chronostratigraphic marker horizons. *Precambrian Research* 172: 301-310.

Kennedy, M.J., B. Runnegar, A.R. Prave, K.-H. Hoffmann, and M.A. Arthur, 1998. Two or four Neoproterozoic glaciations? *Geology* 26: 1059-1063.

Kennedy, M.J., N. Christie-Blick, and L.E. Sohl, 2001. Are Proterozoic cap carbonates and isotopic excursions a record of gas hydrate destabilization following Earth's coldest intervals? *Geology* 29: 443-446.

Kennedy, M.J., D.D. Mrofka, and C.C. von der Borch, 2009. Snowball Earth termination by destabilization of equatorial permafrost methane clathrate. *Nature* 453: 642-645.

Kennett, J.P., and L.D. Stott, 1991. Abrupt deep-sea warming, paleoceanographic changes and benthic extinctions at the end of the Paleocene. *Nature* 353: 225-229.

Kennett, J.P., and L.D. Stott, 1995. Terminal Paleocene mass extinction in the deep sea: Association with global warming. Pp. 94-107 in *Effects of Past Global Change on Life*. Washington, D.C.: National Academy Press.

Kent, D.V., and G. Muttoni, 2008. Equatorial convergence of India and early Cenozoic climate trends. *Proceedings of the National Academy of Sciences USA* 105: 16065-16070.

Kent, D.V., and P.E. Olsen, 1999. Astronomically tuned geomagnetic polarity time scale for the Late Triassic. *Journal of Geophysical Research* 104: 12831-12841.

Kent, D.V., P.E. Olsen, and W.K. Witte, 1995. Late Triassic-Early Jurassic geomagnetic polarity and paleolatitudes from drill cores in the Newark Rift Basin (Eastern North America). *Journal of Geophysical Research* 100: 14965-14998.

Kiehl, J.T., and C.A. Shields, 2005. Climate simulation of the latest Permian: Implications for mass extinction. *Geology* 9: 757-760.

Kienast, M., S.S. Kienast, S.E. Calvert, T.I. Eglinton, G. Mollenhauer, R. François, and A. Mix, 2006. Eastern Pacific cooling and Atlantic overturning circulation during the last deglaciation. *Nature* 443: 846-849.

Kiessling, W., and C. Simpson, 2010. On the potential for ocean acidification to be a general cause of ancient reef crises. *Global Change Biology* 17: 56-67.

Kiessling, W., E. Flügel, and J. Golonka, 1999. Paleoreef maps: Evaluation of a comprehensive database on Phanerozoic reefs. *American Association of Petroleum Geologists Bulletin* 83: 1552-1587.

Kim, S.-J., T.J. Crowley, D.J. Erickson, B. Govindaswamy, P.B. Duffy, and B.Y. Lee, 2008. High-resolution climate simulations of the Last Glacial Maximum. *Climate Dynamics* 31: 1-16.

Kleypas, J.A., J.W. McManus, and L.A.B. Menez, 1999. Environmental limits to coral reef development: Where do we draw the line? *American Zoologist* 39: 146-159.

Kneller, M., and D. Peteet, 1999. Late-glacial to early Holocene climate changes from a Central Appalachian pollen and macrofossil record. *Quaternary Research* 51: 133-147.

Knoll, A.H., R.K. Bambach, D.E. Canfield, and J.P. Grotzinger, 1996. Comparative Earth history and Late Permian mass extinction. *Science* 273: 452-457.

Knutti, R., and G.C. Hegerl, 2008. The equilibrium sensitivity of the Earth's temperature to radiation changes. *Nature Geoscience* 1: 735-743.

Kobashi, R., E.L. Grossman, T.E. Yancey, and D.T. Dockery, 2001. Reevaluation of conflicting Eocene tropical temperature estimates: Molluskan oxygen isotope evidence for warm low latitudes. *Geology* 29: 983-986.

Koch, P.L., W.C. Clyde, R.P. Hepple, M.L. Fogel, S.L. Wing, and J.C. Zachos, 2003. Carbon and oxygen isotope records from paleosols spanning the Paleocene-Eocene boundary, Bighorn Basin, Wyoming. Pp. 49-64 in S.L. Wing, P.D. Gingerich, B. Schmitz, and E. Thomas (eds.), *Causes and Consequences of Globally Warm Climates in the Early Paleogene.* Special Papers 369. Boulder, Colo.: Geological Society of America.

Kominz, M.A., 1984. Oceanic ridge volume and sea-level change—an error analysis. *American Association of Petroleum Geologist Memoirs* 36: 109-127.

Kominz, M.A., and S.F. Pekar. 2001. Oligocene eustasy from two-dimensional sequence stratigraphic backstripping. *Geological Society of America Bulletin* 113: 291-314.

Kopp, R.E., T. Raub, D. Schumann, H. Vali, A.V. Smirnov, and J.L. Kirschvink, 2007. Magnetofossil spike during the Paleocene-Eocene thermal maximum: Ferromagnetic resonance, rock magnetic, and electron microscopy evidence from Ancora, New Jersey, United States. *Palaeoceanography* 22: PA4103, 7 pp.

Korte, C., H.W. Kozur, and J. Veizer, 2005. $\delta^{13}C$ and $\delta^{18}O$ values of Triassic brachiopods and carbonate rocks as proxies for coeval seawater and palaeotemperature. *Palaeogeography, Palaeoclimatology, Palaeoecology* 226: 287-306.

Korty, R.L., K.A. Emanuel, and J.R. Scott, 2008. Tropical cyclone–induced-upper ocean mixing and climate: Application to equable climates. *Journal of Climate* 21: 638-654.

Kowalski, E.A., and D.L. Dilcher, 2003. Warmer paleotemperatures for terrestrial ecosystems. *Proceedings of the National Academy of Sciences USA* 100: 167-170.

Kozdon, R., D.C. Kelly, N.T. Kita, and J.W. Valley, 2009. The "cool tropic paradox": Reassessing aberrant $\delta^{18}O$ in foraminifera by SIMS. *Geochimica et Cosmochimica Acta Supplement* 73: A693

Kraus, M.J., and S. Riggins, 2007. Transient drying during the Paleocene-Eocene Thermal Maximum (PETM): Analysis of paleosols in the Bighorn Basin, Wyoming. *Palaeogeography, Palaeoclimatology, Palaeoecology* 245: 444-461.

Kuiper, K.F., A. Deino, F.J. Hilgen, W. Krijgsman, P.R. Renne, and J.R. Wijbrans, 2008. Synchronizing rock clocks of Earth history. *Science* 320: 500-504.

Kump, L.R., 2002. Reducing uncertainty about carbon dioxide as a climate driver. *Nature* 419: 188-190.

Kump, L.R., 2009. Tipping pointedly cooler. *Science* 323: 1175-1176.

Kump, L.R., and D. Pollard, 2008. Amplification of Cretaceous warmth by biological cloud feedbacks. *Science* 320: 195.

Kump, L.R., A. Pavlov, and M.A. Arthur, 2005. Massive release of hydrogen sulfide to the surface ocean and atmosphere during intervals of ocean anoxia. *Geology* 33: 397-400.

Kump, L.R., T.J. Bralower, and A. Ridgewell, 2009. Ocean acidification in deep time. *Oceanography* 22: 94-107.

Kürschner, W.M., Z. Kvaček, and D.L. Dilcher, 2008. The impact of Miocene atmospheric carbon dioxide fluctuations on climate and the evolution of terrestrial ecosystems. *Proceedings of the National Academy of Sciences USA* 105: 449-453.

Kurtz, A.C., L.R. Kump, M.A. Arthur, J.C. Zachos, and A. Paytan, 2003. Early Cenozoic decoupling of the global carbon and sulfur cycles. *Paleoceanography* 18: 1090, 14 pp.

Langebroek, P.M., A. Paul, and M. Schulz, 2008. Constraining atmospheric CO_2 content during the Middle Miocene Antarctic glaciation using an ice sheet-climate model. *Climate of the Past Discussions* 4: 859-895.

Laskar, J., P. Robutel, F. Joutel, M. Gastineau, A.C.M. Correia, and B. Levrard, 2004. A long-term numerical solution for the insolation quantities of the Earth. *Astronomy and Astrophysics* 428: 261-285.

Lazarus, D.B., B. Kotrc, G. Wulf, and D.N. Schmidt, 2009. Radiolarians decreased silicification as an evolutionary response to reduced Cenozoic ocean silica availability. *Proceedings of the National Academy of Sciences USA* 106: 9333-9338.

Lea, D.W., 2004. The 100,000-yr cycle in tropical SST, greenhouse forcing, and climate sensitivity. *Journal of Climate* 17: 2170-2179.

Lea, D.W., T.A. Mashiotta, and H.J. Spero, 1999. Controls on magnesium and strontium uptake in planktonic foraminifera determined by live culturing. *Geochimica et Cosmochimica Acta* 63: 2369-2379.

Lea, D.W., D.K. Pak, and H.J. Spero, 2000. Climate impact of Late Quaternary equatorial Pacific sea surface temperature variations. *Science* 289: 1719-1724.

Lear, C.H., H. Elderfield, and P.A. Wilson, 2000. Cenozoic deep-sea temperatures and global ice volumes from Mg/Ca in benthic foraminiferal calcite. *Science*, 287: 269-272.

Lear, C.H., Y. Rosenthal, and N. Slowey, 2002. Benthic foraminiferal Mg/Ca-paleothermometry: A revised core-top calibration. *Geochimica et Cosmochimica Acta* 66(19): 3375-3387.

Lear, C.H., T.R. Bailey, P.N. Pearson, H.K. Coxall, and Y. Rosenthal, 2008. Cooling and ice growth across the Eocene-Oligocene transition. *Geology* 36: 251-254.

Leckie, R.M., T.J. Bralower, and R. Cashman, 2002. Oceanic anoxic events and plankton evolution: Biotic response to tectonic forcing during the mid-Cretaceous. *Paleoceanography* 17: 1041, 29 pp.

Lenton, T.M., H. Held, E. Kriegler, J.W. Hall, W. Lucht, S. Rahmstorf, and H.J. Schellnhuber, 2008. Tipping elements in the Earth's climate system. *Proceedings of the National Academy of Sciences USA* 105: 1786-1793.

Lippert, P.C., and Zachos, J.C., 2007. A biogenic origin for anomalous fine-grained magnetic material at the Paleocene-Eocene boundary at Wilson Lake, New Jersey. *Paleoceanography*, 22: PA4104, 8 pp.

Liu, W., and Y. Huang, 2005. Compound-specific D/H ratios and molecular distributions of higher plant leaf waxes as novel paleoenvironmental indicators in the Chinese Loess Plateau. *Organic Geochemistry* 36: 851-860.

Liu, Z., B.L. Otto-Bliesner, F. He, E.C. Brady, R. Tomas, P.U. Clark, A.E. Carlson, J. Lynch-Stieglitz, W. Curry, E. Brook, D. Erickson, R. Jacob, J. Kutzbach, and J. Cheng, 2009a. Transient simulations of last deglaciation with a new mechanism for Bølling-Allerød warming. *Science* 325: 310-314.

Liu, Z., M. Pagani, D. Zinniker, R. DeConto, M. Huber, H. Brinkhuis, S.R. Shah, R.M. Leckie, and A. Pearson, 2009b. Global cooling during the Eocene-Oligocene climate transition. *Science* 323: 1187-1190.

Liu, X., F. Huang, P. Kong, A. Aimin, X. Li, and Y. Ju, 2010. History of ice sheet elevation in East Antarctica: Paleoclimatic implications. *Earth and Planetary Science Letters* 290: 281-288.

Loope, D.B., C.M. Rowe, and R.M. Joeckel, 2001. Annual monsoon rains recorded by Jurassic dunes. *Nature* 412: 64-66.

Loope, D.B., M.B. Steiner, C.M. Rowe, and N. Lancaster, 2004. Tropical westerlies over Pangaean sand seas. *Sedimentology* 51: 315-322.

Lowenstein, T.K., and R.V. Demicco, 2006. Elevated Eocene atmospheric CO_2 and its subsequent decline. *Science* 313: 1928-1928.

Lu, J., G.A. Vecchi, and T. Reichler, 2007. Expansion of the Hadley cell under global warming. *Geophysical Research Letters* 34: L14808, 5 pp.

Ludvigson, G.A., L.A. Gonzalez, R.A. Metzger, B.J. Witzke, R.L. Brenner, A.P. Murillo, and T.S. White, 1998. Meteoric sphaerosiderite lines and their use for paleohydrology and paleoclimatology. *Geology* 26: 1039-1042.

Lunt, D.J., G.L. Foster, A.M. Haywood, and E.J. Stone, 2008. Late Pliocene Greenland glaciation controlled by a decline in atmospheric CO_2 levels. *Nature* 454: 1102-1105.

Lyons, T.W., A.D. Anbar, S. Severmann, C. Scott, and B.C. Gill, 2009. Tracking Euxinia in the ancient ocean: A multiproxy perspective and proterozoic case study. *Annual Review of Earth and Planetary Sciences* 37: 507-534.

Macdonald, F.A., M.D. Schmitz, J.L. Crowley, C.F. Roots, D.S. Jones, A.C. Maloof, J.V. Strauss, P.A. Cohen, D.T. Johnston, and D.P. Schrag, 2010. Calibrating the Cryogenian. *Science* 327: 1241–1243.

McArthur, J.M., T.J. Algeo, B. van de Schootbrugge, Q. Li, and R.J. Howarth, 2008a. Basinal restriction, black shales, Re-Os dating, and the Early Toarcian (Jurassic) oceanic anoxic event. *Paleoceanography* 23: PA4217, 22 pp.

McArthur, J.M., A.S. Cohen, A.L. Coe, D.B. Kemp, R.J. Bailey, and D.G. Smith, 2008b. Discussion on the Late Palaeocene-Early Eocene and Toarcian (Early Jurassic) carbon isotope excursions: A comparison of their time scales, associated environmental change, causes and consequences. *Journal of the Geological Society* 165: 875-880.

McElwain, J.C., 2004. Climate-independent paleoaltimetry using stomatal density in fossil leaves as a proxy for CO_2 partial pressure. *Geology* 32: 1017-1020.

McElwain, J.C., and W.G. Chaloner, 1995. Stomatal density and index of fossil plants track atmospheric carbon dioxide in the Palaeozoic. *Annals of Botany* 76: 389-395.

McElwain, J.C., and D.M. Haworth, 2009. The stomatal-CO_2 proxy: Limitations and advances. *Geochimica et Cosmochimica Acta Supplement* 73: A856-A856.

McElwain, J.C., J. Wade-Murphy, and S.P. Hesselbo, 2005. Changes in carbon dioxide during an oceanic anoxic event linked to intrusion into Gondwana coals. *Nature* 435: 479-482.

Makarieva, A.M., V.G. Gorshkov, and B.L. Li, 2009. Re-calibrating the snake palaeothermometer. *Nature* 460: E2-E3.

Malakoff, D., 1998. Death by suffocation in the Gulf of Mexico. *Science* 281: 190-192.

Marchitto, T.M., W.B Curry, and D.W. Oppo, 2000. Zinc concentrations in benthic foraminifera reflect seawater chemistry. *Paleoceanography* 15: 299-306.

Markwick, P., 1998. Fossil crocodilians as indicators of Late Cretaceous and Cenozoic climates: Implications for using paleontological data in reconstructing palaeoclimate. *Palaeogeography, Palaeoclimatology, Palaeoecology* 137: 207-271.

Markwick, P.J., 2007. The palaeogeographic and paleoclimatic significance of climate proxies for data-model comparisons. Pp. 251-312 in M. Williams, A.M. Haywood, F.J. Gregory, and D.N. Schmidt (eds.), *Deep-Time Perspectives on Climate Change: Marrying the Signal from Computer Models and Biological Proxies.* London: The Micropaleontological Society, Special Publications and The Geological Society.

Marland, G., T.A. Boden, and R.J. Andres, 2002. Global, regional, and national CO_2 emissions. *Trends: A Compendium of Data on Global Change.* Available online at http://cdiac.esd.ornl.gov/trends/trends.htm; accessed July 22, 2010.

Marland, G., R.A. Pielke, Sr., M. Apps, R. Avissar, R.A. Betts, K.J. Davis, P. Frumhoff, S.T. Jackson, L. Joyce, P. Kauppi, K.G. MacDicken, R. Neilson, J.O. Niles, D.S. Niyogi, R.J. Norby, N. Pena, N. Sampson, and Y. Xue, 2003. The climatic implications of land surface change and carbon management, and the implications for climate-change mitigation policy. *Climate Policy* 3: 149-157.

Martrat, B., J.O. Grimalt, N.J. Shackleton, L. de Abreu, M.A. Hutterli, and T.F. Stocker, 2007. Four climate cycles of recurring deep and surface water destabilizations on the Iberian Margin. *Science* 317: 502-507.

Menegatti, P., H. Weissert, R.S. Brown, R.V. Tyson, P. Farrimoud, A. Strasser, and M. Caron, 1998. High-resolution $\delta^{13}C$ stratigraphy through the early Aptian "Livello Selli" of the Alpine Tethys. *Paleoceanography* 13: 530-545.

Meyer, K.M., L.R. Kump, and A. Ridgwell, 2008. Biogeochemical controls on photic-zone euxinia during the end-Permian mass extinction. *Geology* 36: 747-750.

Meyers, S.R., B.B. Sageman, and T.W. Lyons, 2005. Organic carbon burial rate and the molybdenum proxy: Theoretical framework and application to Cenomanian-Turonian oceanic anoxic event 2. *Paleoceanography* 20: PA2002, 19 pp.

Miller, K.G., 2009. Broken greenhouse windows. *Nature Geoscience* 2: 465-466.

Miller, K.G., P.J. Sugarman, J.V. Browning, M.A. Kominz, J.C. Hernandez, R.K. Olsson, J.D. Wright, M.D. Feigenson, and W. Van Sickel, 2003. Late Cretaceous chronology of large, rapid sea-level changes: Glacioeustasy during the greenhouse world. *Geology* 31: 585-588.

Miller, K.G., P.J. Sugarman, J.V. Browning, M.A. Kominz, R.K. Olsson, M.D. Feigenson, and J.C. Hernandez, 2004. Upper Cretaceous sequences and sea level history, New Jersey Coastal Plain. *Geological Society of America Bulletin* 116: 368-393.

Miller, K.G., M.A. Kominz, J.V. Browning, J.D. Wright, G.S. Mountain, M.E. Katz, P.J. Sugarman, B.S. Cramer, N. Christie-Blick, and S.F. Pekar, 2005. The Phanerozoic record of global sea-level change. *Science* 310: 1293-1298.

Miller, G.H., R.B. Alley, J. Brigham-Grette, J.J. Fitzpatrick, L. Polyak, M.C. Serreze, and J.W.C. White, 2010. Arctic amplification: Can the past constrain the future? *Quaternary Science Reviews* 29: 1779-1790.

Montañez, I.P., N.J. Tabor, D. Niemeier, W.A. DiMichele, T.D. Frank, C.R. Fielding, J.L. Isbell, L.P. Birgenheier, and M.C. Rygel, 2007. CO_2-forced climate and vegetation instability during Late Paleozoic deglaciation. *Science* 315: 87-91.

Montoya-Pino, C., S. Weyer, A.D. Anbar, J. Pross, W. Oschmann, B. van de Schootbrugge, and H.W. Arz, 2010. Global enhancement of ocean anoxia during Oceanic Anoxic Event 2: A quantitative approach using U isotopes. *Geology* 38: 315-318.

Moore, E.A. and A.C. Kurtz, 2008. Black carbon in Paleocene-Eocene boundary sediments: A test of biomass combustion as the PETM trigger. *Palaeogeography, Palaeoclimatology, Palaeoecology* 267:147-152.

Morley, R.J., 2000. *Origin and Evolution of Tropical Rain Forests*. New York: John Wiley and Sons, 378 pp.

Moucha, R., A.M. Forte, J.X. Mitrovica, D.B. Rowley, S. Quéré, N.A. Simmons, and S.P. Grand, 2008. Dynamic topography and long-term sea-level variations: There is no such thing as a stable continental platform. *Earth and Planetary Science Letters* 271: 101-108.

Mourik, A.A., J.F. Bijkerk, A. Cascella, S.K. Hüsing, F.J. Hilgen, L.J. Lourens, and E. Turco, 2010. Astronomical tuning of the La Vedova High Cliff section (Ancona, Italy)—Implications of the Middle Miocene climate transition for Mediterranean sapropel formation. *Earth and Planetary Science Letters* 297: 249-261.

Muttoni, G., C. Carcano, E. Garzanti, M. Ghielmi, A. Piccin, R. Pini, S. Rogledi, and D. Sciunnach, 2003. Onset of major Pleistocene glaciations in the Alps. *Geology* 31: 989-992.

Myers, N., and A.H. Knoll, 2001. The biotic crisis and the future of evolution. *Proceedings of the National Academy of Sciences USA* 98: 5389-5392.

Nägler, T.F., A. Eisenhauer, A. Müller, C. Hemleben, and J. Kramers, 2000. The δ^{44}Ca-temperature calibration on fossil and cultured *Globigerinoides sacculifer*: New tool for reconstruction of past sea surface temperatures. *Geochemistry Geophysics Geosystems* 1: 2000GC000091.

Naish, T., R. Powell, R. Levy, G. Wilson, R. Scherer, F. Talarico, L. Krissek, F. Niessen, M. Pompilio, T. Wilson, L. Carter, R. DeConto, P. Huybers, R. McKay, D. Pollard, J. Ross, D. Winter, P. Barrett, G. Browne, R. Cody, E. Cowan, J. Crampton, G. Dunbar, N. Dunbar, F. Florindo, C. Gebhardt, I. Graham, M. Hannah, D. Hansaraj, D. Harwood, D. Helling, S. Henrys, L. Hinnov, G. Kuhn, P. Kyle, A. Läufer, P. Maffioli, D. Magens, K. Mandernack, W. McIntosh, C. Millan, R. Morin, C. Ohneiser, T. Paulsen, D. Persico, I. Raine, J. Reed, C. Riesselman, L. Sagnotti, D. Schmitt, C. Sjunneskog, P. Strong, M. Taviani, S. Vogel, T. Wilch, and T. Williams, 2009. Obliquity-paced Pliocene West Antarctic ice sheet oscillations. *Nature* 458: 322-328.

Nicolo, M.J., G.R. Dickens, C.J. Hollis, and J.C. Zachos, 2007. Multiple early Eocene hyperthermals: Their sedimentary expression on the New Zealand continental margin and in the deep sea. *Geology* 35: 699-702.

Niyogi, D., and Y. Xue, 2006. Soil moisture regulates the biological response of elevated atmospheric CO_2 concentrations in a coupled atmosphere biosphere model. *Global and Planetary Change* 54: 94-108.

Norris, R.D., and U. Röhl, 1999. Carbon cycling and chronology of climate warming during the Palaeocene/Eocene transition. *Nature* 401: 775-778.

Norris, R.D., K.L. Bice, E.A. Magno, and P.A. Wilson, 2002. Jiggling the tropical thermostat in the Cretaceous hothouse. *Geology* 30: 299-302.

NRC (National Research Council), 1995. *Effects of Past Global Change on Life.* Washington, D.C.: National Academy Press, 272 pp.

NRC (National Research Council), 2001. *Grand Challenges in Environmental Sciences.* Washington, D.C.: National Academy Press, 106 pp.

NRC (National Research Council), 2002. *Abrupt Climate Change: Inevitable Surprises.* Washington, D.C.: National Academy Press, 244 pp.

NRC (National Research Council), 2010. *Understanding Climate's Influence on Human Evolution.* Washington, D.C.: The National Academies Press, 128 pp.

Olsen, P.E., 1986. A 40-million-year lake record of Early Mesozoic orbital climatic forcing. *Science* 234: 842-848.

Olsen, P.E., and D.V. Kent, 1996. Milankovitch climate forcing in the tropics of Pangea during the Late Triassic. *Palaeogeography, Palaeoclimatology, Palaeoecology* 122: 1-26.

Olsen, P.E., and D.V. Kent, 1999. Long-period Milankovitch cycles from the Late Triassic and Early Jurassic of eastern North America and their implications for the calibration of the early Mesozoic time scale and the long-term behavior of the planets. *Philosophical Transactions of the Royal Society of London A* 357: 1761-1787.

Olsen, P.E., D.V. Kent, B. Cornet, W.K. Witte, and R.W. Schlische, 1996. High-resolution stratigraphy of the Newark Rift Basin (Early Mesozoic, Eastern North America). *Geological Society of America Bulletin*, 108: 40-77.

Olsen, P.E., C. Koeberl, H. Huber, A. Montanari, S.J. Fowell, M. Et Touhami, and D.V. Kent, 2002. Continental Triassic-Jurassic boundary in central Pangea: Recent progress and discussion of an Ir anomaly. *Geological Society of America Special Papers* 356: 505-522.

Olsen, P.E., D.V. Kent, M. Et-Touhami, and J.H. Puffer, 2003. Cyclo-, magneto-, and biostratigraphic constraints on the duration of the CAMP event and its relationship to the Triassic-Jurassic boundary. Pp. 7-32 in W.E. Hames, J.G. McHone, P.R. Renne, and C. Ruppel (eds.), *The Central Atlantic Magmatic Province: Insights From Fragments of Pangea.* Geophysical Monograph Series 136. Washington, D.C.: American Geophysical Union.

Olsen, P.E., D.V. Kent, and H. Whiteside, 2010. Implications of the Newark Supergroup-based astrochronology and geomagnetic polarity time scale (Newark-APTS) for the tempo and mode of the early diversification of the Dinosauria. *Earth and Environmental Science Transactions of the Royal Society of Edinburgh* 101: 201-229.

Oster, J.L., I.P. Montañez, W.D. Sharp, and K.M. Cooper, 2009. Late Pleistocene California droughts during deglaciation and Arctic warming. *Earth and Planetary Science Letters* 288: 434-443.

Otto-Bliesner, B.L., J. Marshall, J.T. Overpeck, G.H. Miller, A. Hu, and CAPE Last Interglacial Project members, 2006. Simulating Arctic climate warmth and icefield retreat in the last interglaciation. *Science* 311: 1751-1753.

Overpeck, J.T., T. Webb III, and I.C. Prentice, 1985. Quantitative interpretation of fossil pollen spectra: Dissimilarity coefficients and the method of modern analogs. *Quaternary Research* 23: 87-108.

Pagani, M., K.H. Freeman, and M.A. Arthur, 1999. Late Miocene atmospheric CO_2 concentrations and the expansion of C_4 grasses. *Science* 285: 876-879.

Pagani, M., J.C. Zachos, K.H. Freeman, B. Tipple, and S. Bohaty, 2005. Marked decline in atmospheric carbon dioxide concentrations during the Paleogene. *Science* 309: 600-603.

Pagani, M., N. Pedentchouk, M. Huber, A. Sluijs, S. Schouten, H. Brinkhuis, J.S. Sinninghe Damsté, G.R. Dickens, Expedition 302 Scientists, and J. Backman, S. Clemens, T. Cronin, F. Eynaud, J. Gattacceca, M. Jakobsson, R. Jordan, M. Kaminski, J. King, N. Koc, N.C. Martinez, D. McInroy, T.C. Moore, Jr., K. Moran, M. O'Regan, J. Onodera, H. Pälike, B. Rea, D. Rio, T. Sakamoto, D.C. Smith, K.E.K. St. John, I. Suto, N. Suzuki, K. Takahashi, M. Watanabe and M. Yamamoto for Expedition 302 scientists, 2006. Arctic hydrology during global warming at the Palaeocene/Eocene Thermal Maximum. *Nature* 442: 671-675.

Pagani, M., Z. Liu, J. LaRiviere, and A.C. Ravelo, 2010. High Earth-system climate sensitivity determined from Pliocene carbon dioxide concentrations. *Nature Geoscience* 3: 27-30.

Pälike, H., J. Laskar, and N.J. Shackleton, 2004. Geologic constraints on the chaotic diffusion of the solar system. *Geology* 32: 929-932.

Pälike, H., J. Frazier, and J.C. Zachos, 2006a. Extended orbitally forced palaeoclimatic records from the equatorial Atlantic Ceara Rise. *Quaternary Science Reviews* 25: 3138-3149.

Pälike, H., R.D. Norris, J.O. Herrle, P.A. Wilson, H.K. Coxall, C.H. Lear, N.J. Shackleton, A.K. Tripati, and B.S. Wade, 2006b. The heartbeat of the Oligocene climate system. *Science* 314: 1894-1898.

Panchuk, K.M., C. Holmden, and S.A. Leslie, 2006. Local controls on carbon cycling in the Ordovician midcontinent region of North America with implications for carbon isotope secular curves. *Journal of Sedimentary Research* 76: 200-211.

Panchuk, K., A. Ridgwell, and L.R. Kump, 2008. Sedimentary response to Paleocene-Eocene Thermal Maximum carbon release: A model-data comparison. *Geology* 36: 315-318.

Pancost, R.D., D.S. Steart, L. Handley, M.E. Collinson, J.J. Hooker, A.C. Scott, N.V. Grassineau, and I.J. Glasspool, 2007. Increased terrestrial methane cycling at the Palaeocene-Eocene Thermal Maximum. *Nature* 449: 332-335.

Parmesan, C., and G. Yohe, 2003. A globally coherent fingerprint of climate change impacts across natural systems. *Nature* 421: 37-42.

Parrish, J.T., 1998. *Interpreting Pre-Quaternary Climate from the Geologic Record.* New York: Columbia University Press, 348 pp.

Parsons, B., and J.G. Sclater, 1977. An analysis of the variation of the ocean floor bathymetry and heat flow with age. *Journal of Geophysical Research* 82: 803-827.

Passey, B.H., N.E. Levin, T.E. Cerling, F.H. Brown, and J.M. Eiler, 2010. High temperature environments of human evolution in East Africa based on bond ordering in paleosol carbonates. *Proceedings of the National Academy of Sciences USA* 107: 11245-11249.

Payne, J.L., D.J. Lehrmann, J. Wei, M.J. Orchard, D.P. Schrag, and A.H. Knoll, 2004. Large perturbations of the carbon cycle during recovery from the End-Permian extinction. *Science* 305: 506-509.

Pearson, P.N., and M.R. Palmer, 2000. Atmospheric carbon dioxide concentrations over the past 60 million years. *Nature* 406: 695-699.

Pearson, P.N., P.W. Ditchfield, J. Singano, K.G. Harcourt-Brown, C.J. Nicholas, R.K. Olsson, N.J. Shackleton, and M.A. Hall, 2001. Warm tropical sea surface temperatures in the Late Cretaceous and Eocene epochs. *Nature* 413: 481-487.

Pearson, P.N., B.E. van Dongen, C.J. Nicholas, R.D. Pancost, S. Schouten, J.M. Singano, and B.S. Wade, 2007. Stable warm tropical climate through the Eocene epoch. *Geology* 35: 211-214.

Pearson, P.N., I.K. McMillan, B.S. Wade, J.T. Dunkley, H.K. Coxall, P.R. Bown, and C.H. Lear, 2008. Extinction and environmental change across the Eocene-Oligocene boundary in Tanzania. *Geology* 36: 179-182.

Pearson, P.N., G.L. Foster, and B.S. Wade, 2009. Atmospheric carbon dioxide through the Eocene-Oligocene climate transition. *Nature* 461: 1110-1113.

Pedentchouk, N., W. Sumner, B. Tipple, and M. Pagani, 2008. $\delta^{13}C$ and δD compositions of *n*-alkanes from modern angiosperms and conifers: An experimental set up in central Washington State, USA. *Organic Geochemistry* 39: 1066-1071.

Peppe, D.J., D.L. Royer, P. Wilf, and E.A. Kowalski, 2010. Quantification of large uncertainties in fossil leaf paleoaltimetry. *Tectonics* 29: TC3015, 14 pp.

Peteet, D., 2000. Sensitivity and rapidity of vegetational response to abrupt climate change. *Proceedings of the National Academy of Sciences USA* 97: 1359-1361.

Peteet, D.M., and D.H. Mann, 1994. Late-glacial vegetation, tephra, and climatic history of southwestern Kodiak Island, Alaska. *Ecoscience* 1: 255-267.

Peyser, C.E. and C.J. Poulsen, 2008. Controls on permo-carboniferous precipitation over tropical Pangaea: A GCM sensitivity study. *Palaeogeography, Palaeoclimatology, Palaeoecology* 268: 181-192.

Pielke, R.A., Sr., G. Marland, R.A. Betts, T.N. Chase, J.L. Eastman, J.O. Niles, D. Niyogi, and S.W. Running, 2002. The influence of land-use change and landscape dynamics on the climate system—Relevance to climate change policy beyond the radiative effects of greenhouse gases. *Philosophical Transactions of the Royal Society A* 360: 1705-1719.

Pierrehumbert, R.T., 1995. Thermostats, radiator fins and local runaway greenhouse. *Journal of Atmospheric Sciences* 52: 1784-1806.

Pierrehumbert, R.T., 2002. The hydrologic cycle in deep-time climate problems. *Nature* 419: 191-198.

Plint, A.G., and M.A. Kreitner, 2007. Extensive, thin sequences spanning Cretaceous foredeep suggest high-frequency eustatic control: Late Cenomanian, Western Canada foreland basin. *Geology* 35: 735-738.

Polissar, P.J., K.H. Freeman, D.B. Rowley, F.A. McInerney, and B.S. Currie, 2009. Paleoaltimetry of the Tibetan Plateau from D/H ratios of lipid biomarkers. *Earth and Planetary Science Letters* 287: 64-76.

Pollard, D., and R.M. DeConto, 2005. Hysteresis in Cenozoic Antarctic ice-sheet variations. *Global and Planetary Change* 45: 9-21.

Poulsen, C.J., D. Pollard, I. Montañez, and D. Rowley, 2007a. Late Paleozoic tropical climate response to Gondwanan deglaciation. *Geology* 35: 771-774.

Poulsen, C.J., D. Pollard, and T.S. White, 2007b. General circulation model simulation of the $\delta^{18}O$ content of continental precipitation in the middle Cretaceous: A model-proxy comparison. *Geology* 35: 199-202.

Poulsen, C.J., T.A. Ehlers, and N. Insel, 2010. Onset of convective rainfall during gradual Late Miocene rise of the Central Andes. *Science* 328: 490-493.

Prochnow, S.J., L.C. Nordt, S.C. Atchley, and M.R. Hudec, 2006. Multi-proxy paleosol evidence for middle and late Triassic climate trends in eastern Utah. *Palaeogeography, Palaeoclimatology, Palaeoecology* 232: 53–72.

Quade, J., C. Garzione, and J. Eiler, 2007. Paleoelevation reconstruction using pedogenic carbonates. *Reviews in Mineralogy and Geochemistry* 66: 53-87.

Rabalais, N.N., R.E. Turner, R.J. Díaz, and D. Justic, 2009. Global change and eutrophication of coastal waters. *ICES Journal of Marine Science* 66: 1528-1537.

Ramanathan, V., and W.D. Collins, 1991. Thermodynamic regulation of ocean warming by cirrus clouds deduced from observations of the 1987 El Niño. *Nature* 351: 27-32.

Ramezani, J., M.D. Schmitz, V.I. Davydov, S.A. Bowring, W.S. Snyder, and C.J. Northrup, 2007. High-precision U-Pb zircon age constraints on the Carboniferous-Permian boundary in the southern Urals stratotype. *Earth and Planetary Science Letters* 256: 244-257.

Rasbury, E.T., and J.M. Cole, 2009. Directly dating geologic events: U-Pb dating of carbonates. *Reviews of Geophysics* 47: 545-548.

Ravelo, A.C., and M.W. Wara, 2004. The role of the tropical oceans on global climate during a warm period and a major climate transition. *Oceanography* 17: 32-41.

Ravelo, A.C., P.S. Dekens, and M. McCarthy, 2006. Evidence for El Niño-like conditions during the Pliocene. *GSA Today* 16: 4-11.

Ravizza, G., and K.K. Turekian, 1989. Application of the ^{187}Re-^{187}Os system to black shale geochronometry. *Geochimica et Cosmochimica Acta* 53: 3257-3262.

Raymo, M.E., 1994. The initiation of northern hemisphere glaciations. *Paleoceanography* 9: 399-404.

Raynolds, R.G., and K.R. Johnson, 2002. Drilling of the Kiowa Core, Elbert County, Colorado. *Rocky Mountain Geology* 37: 105-109.

Reichow, M.K., M.S. Pringle, A.I. Al'Mukhamedov, M.B. Allen, V.L. Andreichev, M.M. Buslov, C.E. Davies, G.S. Fedoseev, J.G. Fitton, S. Inger, A.Ya. Medvedev, C. Mitchell, V.N. Puchkov, I.Yu. Safonova, R.A. Scott, and A.D. Saunders, 2009. The timing and extent of the eruption of the Siberian Traps large igneous province: Implications for the end-Permian environmental crisis. *Earth and Planetary Science Letters* 277: 9-20.

Retallack, G.J., 2005. Pedogenic carbonate proxies for amount and seasonality of precipitation in paleosols. *Geology* 33: 333-336.

Revelle, R., and H.E. Suess, 1957. Carbon dioxide exchange between atmosphere and ocean and the question of an increase of atmospheric CO_2 during the past decades. *Tellus Series B: Chemical and Physical Meteorology* 9: 18-27.

Ridgwell, A.J., A.J. Watson, and D.E. Archer, 2002. Modeling the response of the oceanic Si inventory to perturbation, and consequences for atmospheric CO_2. *Global Biogeochemical Cycles* 16: 1071, 25 pp.

Ridgwell, A.J., M.J. Kennedy, and K. Caldeira, 2003. Carbonate deposition, climate stability, and neoproterozoic ice ages. *Science* 302: 859-862.

Rind, D., and M. Chandler, 1991. Increased ocean heat transports and warmer climate. *Journal of Geophysical Research* 96: 7437-7461.

Rind, D., R. Goldberg, J. Hansen, C. Rosenzweig, and R. Ruedy, 1990. Potential evapotranspiration and the likelihood of future drought. *Journal of Geophysical Research* 95: 9983-10004.

Roche, D.M., Y. Donnadieu, E. Pucéat, and D. Paillard, 2006. Effect of changes in $\delta^{18}O$ content of the surface ocean on estimated sea surface temperatures in past warm climate. *Paleoceanography* 21: PA2023, 7 pp.

Röhl, U., T. Westerhold, T.J. Bralower, and J.C. Zachos, 2007. On the duration of the Paleocene-Eocene thermal maximum (PETM). *Geochemistry Geophysics Geosystems* 8: Q12002, 13 pp.

Rohling, E.J., E.C. Hopmans, and J.S. Sinninghe Damsté, 2006. Water column dynamics during the last interglacial anoxic event in the Mediterranean (sapropel S5). *Paleoceanography* 21: PA2018, 8 pp.

Rohling, E.J., K. Grant, C. Hemleben, M. Siddal, B.A.A. Hoogakker, M. Bolshaw, and M. Kucera, 2008. High rates of sea-level rise during the last interglacial period. *Nature Geoscience* 1: 38-42.

Rohling, E.J., K. Grant, M. Bolshaw, A.P. Roberts, M. Siddal, Ch. Hemleben, and M. Kucera, 2009. Antarctic temperature and global sea level closely coupled over the past five glacial cycles. *Nature Geoscience* 2: 500-504.

Rosenzweig, C., D. Karoly, M. Vicarelli, P. Neofotis, Q. Wu, G. Casassa, A. Menzel, T.L. Root, N. Estrella, B. Seguin, P. Tryjanowski, C. Liu, S. Rawlins, and A. Imeson, 2008. Attributing physical and biological impacts to anthropogenic climate change. *Nature* 453: 353-358.

Rowley, D.B., 2002. Rate of plate creation and destruction: 180 Ma to present. *Geological Society of America Bulletin* 114: 927-933.

Rowley, D.B., and B.S. Currie, 2006. Palaeo-altimetry of the late Eocene to Miocene Lunpola basin, central Tibet. *Nature* 439: 677-681.

Royer, D.L., R.A. Berner, and D.J. Beerling, 2001a. Phanerozoic atmospheric CO_2 change: Evaluating geochemical and paleobiological approaches. *Earth-Science Reviews* 54: 349-392.

Royer, D.L., S.L. Wing, D.J. Beerling, D.W. Jolley, P.L. Koch, L.J. Hickey, and R.A. Berner, 2001b. Paleobotanical evidence for near present-day levels of atmospheric CO_2 during part of the Tertiary. *Science* 292: 2310-2313.

Royer, D.L., P. Wilf, D.A. Janesko. E.A. Kowalski, and D.L. Dilcher, 2005. Correlations of climate and plant ecology to leaf size and shape: Potential proxies for the fossil record. *American Journal of Botany* 92: 1141-1151.

Royer, D.L., R.A. Berner, and J. Park, 2007. Climate sensitivity constrained by CO_2 concentrations over the past 420 million years. *Nature* 446: 530-532.

Ruddiman, W.F., 2007. *Earth's Climate: Past and Future* (2nd edition). New York: W.H Freeman and Co., 388 pp.

Ruhl, M., M.H.L. Deenen, H.A. Abels, N.R. Bonis, W. Krijgsman, and W.M. Kürschner, 2010. Astronomical constraints on the duration of the early Jurassic Hettangian stage and recovery rates following the end-Triassic mass extinction (St. Audrie's Bay/East Quantoxhead, UK). *Earth and Planetary Science Letters* 295: 262-276.

Russell, A.D., S. Emerson., A.C. Mix, and L.C. Peterson, 1996. The use of foraminiferal U/Ca as an indicator of changes in seawater U content. *Paleoceanography* 11: 649-663.

Russell, A.D., B. Honisch, H.J. Spero, and D. Lea, 2004. Effects of seawater carbonate ion concentration and temperature on shell U, Mg, and Sr in cultured planktonic foraminifera. *Geochimica et Cosmochimica Acta* 68: 4347-4361.

Rygel, M.C., C.R. Fielding, T.D. Frank, and L.P. Birgenheier, 2008. The magnitude of late Paleozoic glacioeustatic fluctuations: A synthesis. *Journal of Sedimentary Research* 78: 500-511.

Sachse, D., J. Radke, and G. Gleixner, 2006. δD values of individual *n*-alkanes from terrestrial plants along a climatic gradient—Implications for the sedimentary biomarker record. *Organic Geochemistry* 37: 469-483.

Sageman, B.B., A.E. Murphy, J.P. Werne, C.A. Ver Straeten, D.J. Hollander, and T.W. Lyons, 2003. A tale of shales: The relative roles of production, decomposition, and dilution in the accumulation of organic-rich strata, Middle-Upper Devonian, Appalachian basin. *Chemical Geology* 195: 229-273.

Sageman, B.B., S.R. Meyers, and M.A. Arthur, 2006. Orbital timescale for the Cenomanian-Turonian boundary stratotype and OAE II, central Colorado, USA. *Geology* 34: 125-128.

Sahagian, D., and A. Proussevitch, 2007. Paleoelevation measurement on the basis of vesicular basalts. *Reviews in Mineralogy and Geochemistry* 66: 195-213.

Sanyal, A., N.G. Hemming, W.S. Broecker, and G.N. Hanson, 1997. Changes in pH in the eastern equatorial Pacific across stage 5-6 boundary based on boron isotopes in foraminifera. *Global Biogeochemical Cycles* 11: 125-133.

Scheibner, C., and R.P. Speijer, 2008. Decline of coral reefs during late Paleocene to early Eocene global warming. *eEarth* 3: 19-26.

Schimmelmann, A., M.D. Lewan, and R.P. Wintsch, 1999. D/H isotope ratios of kerogen, bitumen, oil, and water in hydrous pyrolysis of source rocks containing kerogen types I, II, IIS, and III. *Geochimica et Cosmochimica Acta* 63: 3751-3766.

Schimmelmann, A., A.L. Sessions, and M. Mastalerz, 2006. Hydrogen isotopic (D/H) composition of organic matter during diagenesis and thermal maturation. *Annual Review of Earth and Planetary Sciences* 34: 501-533.

Schiøler, P., J. Andersjerg, O.R. Clausen, G. Dam, K. Dybkjær, L. Hamberg, C. Heilmann-Clausen, E.P. Johannessen, L.E. Kristensen, I. Prince, and J.A. Rasmussen, 2007. Lithostratigraphy of the Palaeogene-Lower Neogene succession of the Danish North Sea. *Geological Survey of Denmark and Greenland Bulletin* 12, 77 pp.

Schmidt, M.W., H.J. Spero, and D.W. Lea, 2004. Links between salinity variation in the Caribbean and North Atlantic thermohaline circulation. *Nature* 428: 160-163.

Schmitz, B., and V. Pujalte, 2007. Abrupt increase in seasonal extreme precipitation at the Paleocene-Eocene boundary. *Geology* 35: 215-218.

Scholle, P.A., and M.A. Arthur, 1980. Carbon isotope fluctuations in Cretaceous pelagic limestones: Potential stratigraphic and petroleum exploration tools. *American Association of Petroleum Geologists Bulletin* 64: 67-87.

Schouten, S., E.C. Hopmans, E. Schefuß, and J.S. Sinninghe Damsté, 2002. Distributional variations in marine crenarchaeotal membrane lipids: A new tool for reconstructing ancient sea water temperatures? *Earth and Planetary Science Letters* 204: 265-274.

Schouten, S., A. Forster, F.E. Panoto, and J.S. Sinninghe Damsté, 2007. Towards calibration of the TEX86 palaeothermometer for tropical sea surface temperatures in ancient greenhouse worlds. *Organic Geochemistry* 38: 1537-1546.

Schrag, D.P., D.J. DePaolo, and F.M. Richter, 1995. Reconstructing past sea surface temperatures: Correcting for diagenesis of bulk marine carbonate. *Geochimica et Cosmochimica Acta* 59: 2265-2278.

Schroeder, P.A., and N.D. Melear, 1999. Stable carbon isotope signatures preserved in authigenic gibbsite from a forested granitic-regolith: Panola Mt., Georgia, USA. *Geoderma* 91: 261-279.

Schult, I., J. Feichter, and W.F. Cooke, 1997. Effect of black carbon and sulfate aerosols on the Global Radiation Budget. *Journal of Geophysical Research* 102: 30107-30117.

Schumann, D., T.D. Raub, R.E. Kopp, J.-L. Guerquin-Kern, T.-D. Wu, I. Rouiller, A.V. Smirnov, S.K. Sears, U. Lücken, S.M. Tikoo, R. Hesse, J.L. Kirschvink, and H. Vali, 2008. Gigantism in unique biogenic magnetite at the Paleocene–Eocene Thermal Maximum. *Proceedings of the National Academy of Sciences USA* 105: 17648-17653.

Scotese, C.R., 2004. A continental drift flipbook. *Journal of Geology* 112: 729-741.

Seki, O., G.L. Foster, D.N. Schmidt, A. Mackensen, K. Kawamura, and R.D. Pancost, 2010. Alkenone and boron-based Pliocene pCO_2 records. *Earth and Planetary Science Letters* 292: 201-211.

Selby, D., and R.A. Creaser, 2005. Direct radiometric dating of the Devonian-Mississippian time-scale boundary using the Re-Os black shale geochronometer. *Geology* 33: 545-548.

Selden, P.A., 2001. Terrestrialization of animals. Pp 71-73 in D.E.G. Briggs and P.R. Crowther (eds.), *Palaeobiology II*. Oxford: Blackwell Science.

Sepkoski, J.J., Jr., 1996. Patterns of Phanerozoic extinction: A perspective from global data bases. Pp. 35-51 in O.H. Walliser (ed.). *Global Events and Event Stratigraphy in the Phanerozoic*. Berlin: Springer.

Sereno, P.C., 1999. The evolution of dinosaurs. *Science* 284: 2137-2147.

Sewall, J.O., and L.C. Sloan, 2006. Come a little bit closer: A high-resolution climate study of the early Paleogene Laramide foreland. *Geology* 34:81-84.

Shackleton, N.J., 1987. Oxygen isotopes, ice volume and sea level. *Quaternary Science Reviews* 6: 183-190.

Shackleton, N., and A. Boersma, 1981. The climate of the Eocene ocean. *Journal of the Geological Society of London* 138: 153-157.

Shackleton, N.J., M.A. Hall, and D. Pate, 1995. Pliocene stable isotope stratigraphy of Site 846. Pp. 337-351 in N.G. Pisias, L.A. Mayer, T.R. Janecek, A. Palmer-Julson, and T.H. van Andel (eds.), *Proceedings of the Ocean Drilling Program, Scientific Results*, Volume 138. College Station, Tex.: Ocean Drilling Program.

Shaffer, G., S.M. Olsen, and J.O.P Pedersen, 2009. Long-term oxygen depletion in response to carbon dioxide emissions from fossil fuels. *Nature Geoscience* 2: 105-109.

Sheehan, P.M., 2001. History of marine biodiversity. *Geological Journal* 36: 231-249.

Sheldon, N.D., 2009. Non-marine records of climate change across the Eocene-Oligocene transition. Pp. 241-248 in C. Koeberl and A. Montanari (eds.), *The Late Eocene Earth—Hothouse, Icehouse, and Impacts*. Special Papers 452. Boulder, Colo.: Geological Society of America.

Sheldon, N.D., and N.J. Tabor, 2009. Quantitative paleoenvironmental and paleoclimatic reconstruction using paleosols. *Earth-Science Reviews* 95: 1-52.

Sheldon, N.D., G.J. Retallack, and S. Tanaka, 2002. Geochemical climofunctions from North American soils and application to paleosols across the Eocene-Oligocene boundary in Oregon. *Journal of Geology* 110: 687-696.

Shevenell, A.E., J.P. Kennett, and D.W. Lea, 2004. Middle Miocene Southern Ocean cooling and Antarctic cryosphere expansion. *Science* 305: 1766-1770.

Shu, D.-G., H.-L. Luo, S. Conway Morris, X.-L. Zhang, S.-X. Hu, L. Chen, J. Han, Y. Li, and L.-Z. Chen, 1999. Lower Cambrian vertebrates from South China. *Nature* 402: 42-46.

Shukla, J., and Y. Mintz, 1982. Influence of land-surface evapotranspiration on the Earth's climate. *Science* 215: 1498-1501.

Shukla, S.P., M.A. Chandler, J. Jonas, L.E. Sohl, K. Mankoff, and H. Dowsett, 2009. Impact of a permanent El Niño (El Padre) and Indian Ocean dipole in warm Pliocene climates. *Paleoceanography* 24: PA2221. doi:10.1029/2008PA001682.

Siegenthaler, U., T.F. Stocker, E. Monnin, D. Lüthi, J. Schwander, B. Stauffer, D. Raynaud, J.-M. Barnola, H. Fischer, V. Masson-Delmotte, and J. Jouzel, 2005. Stable carbon cycle-climate relationship during the late Pleistocene. *Science* 310: 1313-1317.

Sigman, D.M., and E.A. Boyle, 2000. Glacial/interglacial variations in atmospheric carbon dioxide. *Nature* 407: 859-869.

Sinha, A., K.G. Cannariato, L.D. Stott, H.-C. Li, C.-F. You, H. Cheng, R.L. Edwards, and I.B. Singh, 2005. Variability of Southwest Indian summer monsoon precipitation during the Bølling-Ållerød. *Geology* 33: 813-816.

Slingo, J., K. Bates, N. Nikiforakis, M. Piggott, M. Roberts, L. Shaffrey, I. Stevens, P.L. Vidale, and H. Weller, 2009. Developing the next-generation climate system models: Challenges and achievements. *Philosophical Transactions of the Royal Society A* 367: 815-831.

Sloan, L.C., and Pollard, D., 1998. Polar stratospheric clouds: A high latitude winter warming mechanism in an ancient greenhouse world. *Geophysical Research Letters* 25: 3517-3520.

Sluijs, A., S. Schouten, M. Pagani, M. Woltering, H. Brinkhuis, J.S. Sinninghe Damsté, G.R. Dickens, M. Huber, G.-J. Reichart, R. Stein, J. Matthiessen, L.J. Lourens, N. Pedentchouk, J. Backman, and K. Moran, 2006. Subtropical Arctic Ocean temperatures during the Palaeocene/Eocene thermal maximum. *Nature* 441: 610-613.

Smith, F.A., and K.H. Freeman, 2006. Influence of physiology and climate on δD of leaf wax *n*-alkanes from C_3 and C_4 grasses. *Geochimica et Cosmochimica Acta* 70: 1172-1187.

Sniderman, J.M.K., 2009. Biased reptilian palaeothermometer? *Nature* 460: E1-E2.

Soreghan, G.S., M.J. Soreghan, C.J. Poulsen, R.A. Young, C.F. Eble, D.E. Sweet, and O.C. Davogustto, 2008. Anomalous cold in the Pangaean tropics. *Geology* 36: 659-662.

Spero, H.J., J. Bijma, D.W. Lea, and B.E. Bemis, 1997. Effect of seawater carbonate concentration on foraminiferal carbon and oxygen isotopes. *Nature* 390: 497-500.

Spicer, R.A., A. Ahlberg, A.B. Herman, C.-C. Hofmann, M. Raikevich, P.J. Valdes, and P.J. Markwick, 2008. The Late Cretaceous continental interior of Siberia: A challenge for climate models. *Earth and Planetary Science Letters* 267: 228-235.

Spivack, A.J., and C.-F. You, 1997. Boron isotopic geochemistry of carbonates and pore waters, Ocean Drilling Program Site 851. *Earth and Planetary Science Letters* 152: 113-122.

Sriver, R.L., and M. Huber, 2007. Observational evidence for an ocean heat pump induced by tropical cyclones. *Nature* 447: 577-580.

Stanford, J.D., E.J. Rohling, S.E. Hunter, A.P. Roberts, S.O. Rasmussen, E. Bard, J. McManus, and R.G. Fairbanks, 2006. Timing of meltwater pulse 1a and climate responses to meltwater injections. *Paleoceanography* 21: PA4103, 9 pp.

Stap, L., A. Sluijs, E. Thomas, and L. Lourens, 2009. Patterns and magnitude of deep sea carbonate dissolution during Eocene Thermal Maximum 2 and H2, Walvis Ridge, southeastern Atlantic Ocean, *Paleoceanography* 24: PA1211.

Stein, C.A., and S. Stein, 1992. A model for the global variation in oceanic depth and heat flow with lithospheric age. *Nature* 359: 123-129.

Stein, C.A., and S. Stein, 1997. Estimation of lateral hydrothermal flow distance from spatial variations in oceanic heat flow. *Geophysical Research Letters* 24: 2323-2326.

Stern L.A., C. Page Chamberlain, R.C. Reynolds, and G.D. Johnson, 1997. Oxygen isotope evidence of climate change from pedogenic clay minerals in the Himalayan molasse. *Geochimica et Cosmochimica Acta* 61: 731-744.

Sternberg, L.S.L., 1988. D/H ratios of environmental water recorded by D/H ratios of plant lipids. *Nature* 333: 59-61.

Stiles, C.A., C.I. Mora, and S.G. Driese, 2001. Pedogenic iron-manganese nodules in vertisols: A new proxy for paleoprecipitation? *Geology* 29: 943-946.

Stoll, H.M., and D.P. Schrag, 1996. Evidence for glacial control of rapid sea level changes in the Early Cretaceous. *Science* 272: 1771-1774.

Stoll, H.M., and D.P. Schrag, 2000. High-resolution stable isotope records from the Upper Cretaceous rocks of Italy and Spain: Glacial episodes in a greenhouse planet? *Geological Society of America Bulletin* 112: 308-319.

Storey, M., R.A. Duncan, and C.C. Swisher, III, 2007. Paleocene-Eocene Thermal Maximum and the opening of the Northeast Atlantic. *Science* 316: 587-589.

Stramma, L., G.C. Johnson, J. Sprintall, and V. Mohrholz, 2008. Expanding oxygen-minimum zones in the tropical oceans. *Science* 320: 655-658.

Strasser, A., F.J. Hilgen, and P.H. Heckel, 2006. Cyclostratigraphy—Concepts, definitions, and applications. *Newsletters on Stratigraphy* 42: 75-114.

Sun, D.-Z., and Z. Liu, 1996. Dynamic ocean-atmosphere coupling: A thermostat for the tropics. *Science* 272: 1148-1150.

Svensen, H., S. Planke, A. Malthe-Sørenssen, B. Jamtveit, R. Myklebust, T. Rasmussen, and S.R. Sebastian, 2004. Release of methane from a volcanic basin as a mechanism for initial Eocene global warming. *Nature* 429: 542-545.

Svensen, H., S. Planke, L. Chevallier, A. Malthe-Sørenssen, F. Corfu, and B. Jamtvelt, 2007. Hydrothermal venting of greenhouse gases triggering Early Jurassic global warming. *Earth and Planetary Science Letters* 256: 554-566.

Svensen, H., S. Planke, A.G. Polozov, N. Schmidbauer, F. Corfu, Y.Y. Podladchikov, and B. Jamtveit, 2009. Siberian gas venting and the end-Permian environmental crisis. *Earth and Planetary Science Letters* 277: 490-500.

Tabor, N.J., 2007. Permo-Pennsylvanian palaeotemperatures from Fe-Oxide and phyllosilicate $\delta^{18}O$ values. *Earth and Planetary Science Letters* 253: 159-171.

Tabor, N.J., and I.P. Montañez, 2005. Oxygen and hydrogen isotope compositions of Permian pedogenic phyllosilicates: Development of modern surface domain arrays and implications for paleotemperature reconstructions. *Palaeogeography, Palaeoclimatology, Palaeoecology* 223: 127-146.

Tabor, N.J., and C.J. Yapp C.J., 2005. Incremental vacuum dehydration-decarbonation experiments on a natural gibbsite (α-Al(OH3)): CO_2 abundance and $\delta^{13}C$ values. *Geochimica et Cosmochimica Acta* 69: 519-527.

Tabor, N.J., I.P. Montañez, and R.J. Southard, 2002. Mineralogical and stable isotopic analysis of pedogenic proxies in Permo-Pennsylvanian paleosols: Implications for paleoclimate & paleoatmospheric circulation. *Geochemica et Cosmochimica Acta* 66: 3093-3107.

Tauxe, L., and D.V. Kent, 2004. A simplified statistical model for the geomagnetic field and the detection of shallow bias in paleomagnetic inclinations: Was the ancient magnetic field dipolar? Pp. 101-115 in J.E.T. Channell, D.V. Kent, W. Lowrie, and J. Meert (eds.), *Timescales of the Paleomagnetic Field*. AGU Geophysical Monograph 145.

Taylor, K.C., G.W. Lamorey, G.A. Doyle, R.B. Alley, P.M. Grootes, P.A. Mayewski, J.W.C. White, and L.K. Barlow, 1993. The flickering switch of late Pleistocene climate change. *Nature* 361: 432-436.

Tejada, M.L.G., J.J. Mahoney, C.R. Neal, R.A. Duncan, and M.G. Petterson, 2002. Basement geochemistry and geochronology of Central Malaita, Solomon Islands, with implications for the origin and evolution of the Ontong Java Plateau. *Journal of Petrology* 43: 449-484.

Thomas, E., 2007. Cenozoic mass extinctions in the deep sea: What perturbs the largest habitat on Earth. Pp. 1–23 *in* S. Monechi, R. Coccioni, and M.R. Rampino, (eds.), Large Ecosystem Perturbations: Causes and Consequences. Special Papers 424. Boulder, Colo.: Geological Society of America.

Thomas, E., and N.J. Shackleton, 1996. The Paleocene-Eocene benthic foraminiferal extinction and stable isotope anomalies. Pp. 401–441 in R.W.O'B. Knox, R.M. Corfield, and R.E. Dunay (eds.), *Correlation of the Early Paleogene in Northwest Europe*. Geological Society of London, Special Publication 101.

Thomas, E., J. C. Zachos, and T. J. Bralower, 2000. Deep-sea environments on a warm Earth: Latest Paleocene-early Eocene. Pp.132-160 in B. Huber, K. MacLeod, and S. Wing (eds.), *Warm Climates in Earth History*. Cambridge, UK: Cambridge University Press.

Thompson, W.G., and S.L. Goldstein, 2005. Open-system coral ages reveal persistent suborbital sea-level cycles. *Science* 308: 401- 404.

Thrasher, B.L., and L.C. Sloan, 2009. Carbon dioxide and the early Eocene climate of western North America. *Geology* 37(9): 807-810.

Tong, J.A., Y. You, R.D. Müller, and M. Seton, 2009. Climate model sensitivity to atmospheric CO_2 concentrations for the middle Miocene. *Global and Planetary Change* 67: 129-140.

Tripati, A.K., R.A. Eagle, N. Thiagarajan, A.C. Gagnon, H. Bauch, P.R. Halloran, and J.M. Eiler, 2010. ^{13}C-^{18}O isotope signatures and "clumped isotope" thermometry in foraminifera and coccoliths. *Geochimica et Cosmochimica Acta* 74: 5697-5717.

Trotter, J.A., I.S. Williams, C.R. Barnes, C. Lecuyer, and R.S. Nicoll, 2008. Did cooling oceans trigger Ordovician biodiversification? Evidence from conodont thermometry. *Science* 321: 550-554.

Tsandev, I., and C.P. Slomp, 2009. Modeling phosphorus cycling and carbon burial during Cretaceous Oceanic Anoxic Events. *Earth and Planetary Science Letters* 286: 71-79.

Turgeon, S.C., and R.A. Creaser, 2008. Cretaceous oceanic anoxic event 2 triggered by a massive magmatic episode. *Nature* 454: 323-326.

Ufnar, D.F., L.A. Gonzalez, G.A. Ludvigson, R.L. Brenner, and B.J. Witzke, 2002. The mid-Cretaceous water bearer: Isotope mass balance quantification of the Albian hydrologic cycle. *Palaeogeography, Palaeoclimatology, Palaeoecology* 188: 51-71.

Ufnar, D.F., L.A. Gonzalez, G.A. Ludvigson, R.L. Brenner, and B.J. Witzke, B.J., 2004. Evidence for increased latent heat transport during the Cretaceous (Albian) greenhouse warming. *Geology* 32: 1049-1052.

Ufnar, D.F., G.A. Ludvigson, L. Gonzalez, and D.R. Grocke, 2008. Precipitation rates and atmospheric heat transport during the Cenomanian greenhouse warming in North America: Estimates from a stable isotope mass-balance model. *Palaeogeography, Palaeoclimatology, Palaeoecology* 266: 28-38.

Van der Voo, R., 1993. Paleomagnetism of the Atlantic, Tethys, and Iapetus Oceans. Cambridge, UK: Cambridge University Press, 411 pp.

van de Wal, R.S.W., W. Boot, M.R. van den Broeke, C.J.P.P. Smeets, C.H. Reijmer, J.J.A. Donker, and J. Oerlemans, 2008. Large and rapid melt-induced velocity changes in the ablation zone of the Greenland Ice Sheet. *Science* 321: 111-113.

Vecchi, G.A., K.L. Swanson, and B.J. Soden, 2008. Whither hurricane activity? *Science* 322: 687-689.

Veizer, J., D. Ala, K. Azmy, P. Bruckschen, D. Buhl, F. Bruhn, G.A.F. Carden, A. Diener, S. Ebneth, Y. Godderis, T. Jasper, C. Korte, F. Pawellek, O.G. Podlaha, and H. Strauss, 1999. $^{87}Sr/^{86}Sr$, $\delta^{13}C$, and $\delta^{18}O$ evolution of Phanerozoic seawater. *Chemical Geology* 161: 59-88.

Vitali, F., F.J. Longstaffe, P.J. McCarthy, A.G. Plint, and W.G.E. Caldwell, 2002. Stable isotopic investigation of clay minerals and pedogenesis in an interfluve paleosol from the Cenomanian Dunvegan Formation, N.E. British Columbia, Canada. *Chemical Geology* 192: 269-287.

Veron, J.E.N., 2008. Mass extinctions and ocean acidification: Biological constraints on geological dilemmas. *Coral Reefs* 27: 459-472.

Wagner, J.D.M., J.E. Cole, J.W. Beck, P.J. Patchett, G.M. Henderson, and H.R. Barnett, 2010. Moisture variability in the southwestern United States linked to abrupt glacial climate change. *Nature Geoscience* 3: 110-113.

Wagner, T., S. Damsté, S. Japp, P. Hofmann, and B. Beckmann, 2004. Euxinia and primary production in Late Cretaceous eastern equatorial Atlantic surface waters fostered orbitally driven formation of marine black shales. *Paleoceanography* 19: PA3009, 13 pp.

Walker, J.C.G., and J.F. Kasting, 1992. Effect of forest and fuel conservation on future levels of atmospheric carbon dioxide. *Palaeogeography, Palaeoclimatology, Palaeoecology* 97: 151-189.

Wang, Y.J., H. Cheng, R.L. Edwards, Z.S. An, J.Y. Wu, C.-C. Shen, and J.A. Dorale, 2001. A high-resolution absolute-dated Late Pleistocene monsoon record from Hulu Cave, China. *Science* 294: 2345-2348.

Wang, Y., H. Cheng, R.L. Edwards, Y. He, X. Kong, Z. An, J. Wu, M.J. Kelly, C.A. Dykoski, and X. Li, 2005. The Holocene Asian monsoon: Links to solar changes and North Atlantic climate. *Science* 308: 854-857.

Wang, Z., A. Cohen, G. Gaetani, R. Gabitov, and S. Hart, 2008. Mg isotope fractionation in corals: Developing a promising paleothermometer. Abstract, *11th International Coral Reef Symposium 2008*, Fort Lauderdale, Florida.

Wara, M.W., A.C. Ravelo, and M.L. Delaney, 2005. Permanent El Niño-like conditions during the Pliocene warm period. *Science* 309: 758-761.

Wardlaw, B.R., and T.M. Quinn, 1991. The record of Pliocene sea-level change at Enewetak Atoll. *Quarternary Science Reviews* 10: 247-258.

Weiss, J.L., J.T. Overpeck, and B. Stein, 2011. Implications of recent sea level rise science for low-elevation areas in coastal cities of the U.S.A. *Climatic Change* 105(3-4): 635-645.

Weissert, H., 1989. C-isotope stratigraphy, a monitor of paleoenvironmental change: A case study from the early Cretaceous. *Surveys in Geophysics* 10: 1-61.

Weissert, H., and E. Erba, 2004. Volcanism, CO_2 and palaeoclimate: A Late Jurassic-Early Cretaceous carbon and oxygen isotope record. *Journal of the Geological Society of London* 161: 695-702.

Wentz, F.J., L. Ricciardulli, K. Hilburn, and C. Mears, 2007. How much more rain will global warming bring? *Science* 317: 233-235.

Westerhold, T., U. Röhl, I. Raffi, E. Fornaciari, S. Monechi, V. Reale, J. Bowles, and H.F. Evans, 2008. Astronomical calibration of the Paleocene time. *Palaeogeography, Palaeoclimatology, Palaeoecology* 257: 377-403.

White, J.W.C., P. Ciais, R.A. Figge, R. Kenny, and V. Markgraf, 1994. A high-resolution record of atmospheric CO_2 content from carbon isotopes in peat. *Nature* 367: 153-156.

White, T.S., L.A. Gonzalez, G.A. Ludvigson, and C. Poulsen, 2001. The mid-Cretaceous hydrologic cycle of North America. *Geology* 29: 363-366.

Whiteside, J.H., P.E. Olsen, T.I. Eglinton, M.E. Brookfield, and R.N. Sambrotto, 2010. Compound-specific carbon isotopes from Earth's largest flood basalt province directly link eruptions to the end-Triassic mass extinction. *Proceedings of the National Academy of Sciences USA* 107: 6721-6725.

Wignall, P.B., 2007. The end-Permian mass extinction—how bad did it get? *Geobiology* 5: 303–309.

Wilf, P., 1997. When are leaves good thermometers? A new case for leaf margin analysis. *Paleobiology* 23: 373 390.

Wilson, P.A., and R.D. Norris, 2001. Warm tropical ocean surface and global anoxia during the mid-Cretaceous period. *Nature* 412: 425-429.

Windley, B.F., 1995. *The Evolving Continents.* New York: Wiley, 544 pp.

Wing, S.L., P.D. Gingerich, B. Schmitz, and E. Thomas (eds.), 2003. *Causes and Consequences of Globally Warm Climates in the Early Paleogene*. Special Papers 369. Boulder, Colo.: Geological Society of America.

Wing, S.L., G.J. Harrington, F.A. Smith, J.L. Bloch, D.M. Boyer, and K.H. Freeman, 2005. Transient floral change and rapid global warming at the Paleocene-Eocene boundary. *Science* 310: 993-996.

Winguth, A.M.E., and E. Maier-Reimer, 2005. Causes of marine productivity and oxygen changes associated with the Permian-Triassic boundary: A reevalution with ocean general circulation models. *Marine Geology* 217: 283-304.

Woods, A.D., D.J. Bottjer, and F.A. Corsetti, 2007. Deep-water seafloor calcite precipitates from east-central California: Sedimentology and paleobiological significance. *Palaeogeography, Palaeoclimatology, Palaeoecology* 252: 281-290.

Woodward, F.I., 1987. Stomatal numbers are sensitive to increases in CO_2 from pre-industrial levels. *Nature* 327: 617-618.

Wolfe, J.A., 1993. A method of obtaining climatic parameters from leaf assemblages. *U.S. Geological Survey Bulletin* 2040: 1-71.

Wolfe, J.A., 1995. Paleoclimatic estimates from Tertiary leaf assemblages. *Annual Reviews in Earth and Planetary Science* 23: 119-142.

Xie, S.-P., C. Deser, G.A. Vecchi, J. Ma, H. Teng, and A.T. Wittenberg, 2010. Global warming pattern formation: Sea surface temperature and rainfall. *Journal of Climate* 23: 966-986.

Yapp, C.J., 2000. Climatic implications of surface domains in arrays of δD and $\delta^{18}O$ from hydroxyl minerals: Goethite as an example. *Geochimica et Cosmochimica Acta* 64: 2009-2025.

Yapp, C.J., 2004. $Fe(CO_3)OH$ in goethite from a mid-latitude North American Oxisol: Estimate of atmospheric CO_2 concentration in the Early Eocene "climatic optimum." *Geochimica et Cosmochimica Acta* 68: 935-947.

Yapp, C.J., and H. Poths, 1992. Ancient atmospheric CO_2 pressures inferred from natural goethites. *Nature* 355: 342-344.

Yuan, D., H. Cheng, R.L. Edwards, C.A. Dykoski, M.J. Kelly, M. Zhang, J. Qing, Y. Lin, Y. Wang, J. Wu, J.A. Dorale, Z. An, and Y. Cai, 2004. Timing, duration, and transitions of the Last Interglacial Asian monsoon. *Science* 304: 575-578.

Zachos, J.C., and L.R. Kump, 2005. Carbon cycle feedbacks and the initiation of Antarctic glaciation in the earliest Oligocene. *Global and Planetary Change* 47: 51-66.

Zachos, J.C., T.M. Quinn, and S. Salamy, 1996. High resolution (10^4 yr) deep-sea foraminiferal stable isotope records of the earliest Oligocene climate transition. *Paleoceanography* 9: 353-387.

Zachos, J., M. Pagani, L. Sloan, E. Thomas, and K. Billups, 2001a. Trends, rhythms, and aberrations in global climate 65 Ma to present. *Science* 292: 686-693.

Zachos, J.C., N.J. Shackleton, J.S. Revenaugh, H. Pälike, and B.P. Flower, 2001b. Climate response to orbital forcing across the Oligocene-Miocene boundary. *Science* 292: 274-278

Zachos, J.C., M.W. Wara, S. Bohaty, M.L. Delaney, M.R. Petrizzo, A. Brill, T.J. Bralower, and I. Premoli-Silva, 2003. A transient rise in tropical sea surface temperature during the Paleocene-Eocene Thermal Maximum. *Science* 302: 1551-1554.

Zachos, J.C., U. Röhl, S.A. Schellenberg, A. Sluijs, D.A. Hodell, D.C. Kelly, E. Thomas, M. Nicolo, I. Raffi, L.J. Lourens, H. McCarren, and D. Kroon, 2005. Rapid acidification of the ocean during the Paleocene-Eocene Thermal Maximum. *Science* 308: 1611-1615.

Zachos, J.C., S. Schouten, S. Bohaty, T. Quattlebaum, A. Sluijs, H. Brinkhuis, S.J. Gibbs, and T.J. Bralower, 2006. Extreme warming of mid-latitude coastal ocean during the Paleocene-Eocene Thermal Maximum: Inferences from TEX_{86} and isotope data. *Geology* 34: 737-740.

Zachos, J.C., G.R. Dickens, and R.E. Zeebe, 2008. An early Cenozoic perspective on greenhouse warming and carbon-cycle dynamics. *Nature* 451: 279-283.

Zanazzi, A., and M.J. Kohn, 2008. Ecology and physiology of White River mammals based on stable isotope ratios of teeth. *Palaeogeography, Palaeoclimatology, Palaeoecology* 257: 22-37.

Zeebe, R.E., 1999. An explanation of the effect of seawater carbonate concentration on foraminiferal oxygen isotopes. *Geochimica et Cosmochimica Acta* 63: 2001-2007.

Zeebe, R.E., and K. Caldeira, 2008. Close mass balance of long-term carbon fluxes from ice-core CO_2 and ocean chemistry records. *Nature Geoscience* 1: 312-315.

Zeebe, R.E., and J.C. Zachos, 2007. Reversed deep-sea carbonate ion basin gradient during Paleocene-Eocene thermal maximum. *Paleoceanography* 22: PA3201.

Zeebe, R.E., J.C. Zachos, K. Caldeira, and T. Tyrrell, 2008. Carbon emissions and acidification. *Science* 321: 51-52.

Zeebe, R.E., J.C. Zachos, and G.R. Dickens, 2009. Carbon dioxide forcing alone insufficient to explain Palaeocene-Eocene Thermal Maximum warming. *Nature Geoscience* 2: 576-580.

Zhang, X., W. Lin, and M. Zhang, 2007. Toward understanding the double Intertropical Convergence Zone pathology in coupled ocean-atmosphere general circulation models. *Journal of Geophysical Research* 112: D12102, 7 pp.

Zhang, P., H. Cheng, R.L. Edwards, F. Chen, Y. Wang, X. Yang, J. Liu, M. Tan, X. Wang, J. Liu, C. An, Z. Dai, J. Zhou, D. Zhang, J. Jia, L. Jin, and K.R. Johnson, 2008. A test of climate, sun, and culture relationships from an 1810-year Chinese cave record. *Science* 322: 940-942.

Zhou, J., C.J. Poulsen, D. Pollard, and T.S. White, 2008. Simulation of modern and middle Cretaceous marine $\delta^{18}O$ with an ocean-atmosphere general circulation model. *Paleoceanography* 23: PA3223, 11 pp.

Ziegler, A.M., C.R. Scotese, and S.F. Barrett, 1983. Mesozoic and Cenozoic paleogeographic maps. Pp. 240-252 in P. Broche, and J. Sundermann (eds.), *Tidal Friction and the Earth's Rotation*. II. Berlin: Springer-Verlag.

Ziegler, A., G. Eshel, P.M. Rees, T. Rothfus, D. Rowley, and D. Sunderlin, 2003. Tracing the tropics across land and sea: Permian to present. *Lethaia* 36: 227-254.

Appendixes

Appendix A

Committee Biographical Sketches

Isabel P. Montañez *(Chair)* is a professor in the Department of Geology at the University of California, Davis. Dr. Montañez is a field geologist and geochemist whose research focuses on the sedimentary archive of paleoatmospheric composition and paleoclimatic conditions, in particular reconstructing records of greenhouse gas-climate linkages during periods of major climate transitions. Her past work has involved study of marine and terrestrial successions of Cambrian through Pleistocene age. Dr. Montañez is a fellow of the Geological Society of America. She received her Ph.D. in geology from Virginia Polytechnic Institute and State University.

Thomas J. Algeo is a professor in the Department of Geology at the University of Cincinnati. Prior to joining the faculty at Cincinnati, Dr. Algeo worked as a petroleum geologist at both Exxon and Amoco exploration companies. His research focuses on the processes driving long-term development of Earth's ocean-atmosphere-biosphere systems, using stratigraphic and geochemical proxies from marine carbonates and black shales. Dr. Algeo received his Ph.D. in geological sciences from the University of Michigan.

Mark A. Chandler is an associate research scientist at the Center for Climate Systems Research at Columbia University. Dr. Chandler's primary research involves the use of global climate models to analyze Earth's past climates, from previous times of global warming to snowball Earth episodes. His other major research focus is on improving the usability and accessibility of computer three-dimensional climate models. Dr. Chandler directs the Educational Global Climate Modeling Project, which develops, distributes, and supports a fully functional version of the National Aeronautics and Space Administration/Goddard Institute for Space Studies General Circulation Model Model II for use in precollege and university-level science courses. He received his Ph.D. in geological sciences from Columbia University.

Kirk R. Johnson is vice president of Research and Collections and Chief Curator at the Denver Museum of Nature and Science. Dr. Johnson's research interests span paleobotany, paleoecology, biogeography, geochronology, and biostratigraphy with a particular focus on the Cretaceous to Eocene period. As well as his research publications, Dr. Johnson is the lead author of several popular science books and he has appeared on numerous television programs to popularize geoscience concepts. Dr. Johnson received his Ph.D. in geology and paleobotany from Yale University, and he is a fellow of the Geological Society of America.

Martin J. Kennedy is a professor in the School of Earth and Environmental Sciences at the University of Adelaide, Australia. Previously, he was a professor in the Department of Earth Sciences at the University of California, Riverside, and before then was a senior research geologist at Exxon Production Research Co. His research interests are focused on paleoceanographic and paleoclimate events recorded in the stratigraphic record, using sedimentological and geochemical data integrated with high-resolution sequence and isotope stratigraphic techniques to understand controls of the ancient carbon cycle and biogeochemical feedbacks within the biosphere. Dr. Kennedy received his Ph.D. from the University of Adelaide, Australia.

Dennis V. Kent (NAS) is a professor in the Department of Earth and Planetary Sciences at Rutgers University and an adjunct senior research scientist at Lamont-Doherty Earth Observatory. Dr. Kent's research interests focus on the use of Cenozoic and Mesozoic magnetostratigraphy and geomagnetic polarity timescales to address geological problems, including paleoclimatology and paleogeography. Dr. Kent is a fellow of the American Association for the Advancement of Science, the American Geophysical Union, and the Geological Society of America. He received his Ph.D. in marine geology and geophysics from Columbia University.

Jeffrey T. Kiehl is a senior scientist in the Climate Change Research Section of the National Center for Atmospheric Research (NCAR) in Boulder, Colorado. Dr. Kiehl's current research focuses on using climate modeling to understand Earth's warm greenhouse climates for deep-time periods ranging between 300 and 50 million years ago, and on understanding climate feedback processes in Earth's climate system. Dr. Kiehl has served on the National Research Council Committee on Global Change Research and Climate Research Committee, and he was a contributing author for the Intergovernmental Panel on Climate Change Third Assessment Report. He received his Ph.D. in atmospheric science from the State University of New York, Albany.

Lee R. Kump is a professor of geosciences at the Pennsylvania State University and is also currently associate director of the Earth System Evolution Program at the Canadian Institute for Advanced Research. Dr. Kump's research focuses on the long-term evolution of the oceans and atmosphere and the dynamic coupling between global climate and biogeochemical cycles. He is a fellow of the Geological Society of America and the Geological Society of London. Dr. Kump received his Ph.D. in marine sciences from the University of South Florida.

Richard D. Norris is professor of paleobiology at the Scripps Institution of Oceanography of the University of California, San Diego. Dr. Norris's research interests focus on the use of biogeochemical and paleoceanographic data to understand Earth-ocean-biosphere linkages, with particular emphasis on Cretaceous and Paleogene warm climates and the Cretaceous Thermal Maximum. Dr. Norris received his Ph.D. from Harvard University, and he is a fellow of the Geological Society of America.

A. Christina Ravelo is a professor of ocean sciences in the Department of Ocean Sciences at the University of California, Santa Cruz (UCSC). She is also director of the Santa Cruz branch of the Institute of Geophysics and Planetary Physics (IGPP) at UCSC. Previously, she was chair of the U.S. Science Advisory Committee for Scientific Ocean Drilling and director of the IGPP's Center for the Dynamics and Evolution of the Land-Sea Interface. Dr. Ravelo's research interests are focused on understanding Cenozoic paleoclimates and paleoceanography using stable isotope geochemistry. She received her Ph.D. in geological sciences from Columbia University.

Karl K. Turekian (NAS) is the Sterling Professor of Geology and Geophysics at Yale University. Dr. Turekian's research focuses on the use of radioactive and radiogenic nuclides for deciphering the environmental history of Earth. He received his Ph.D. in geochemistry from Columbia University and has served on the faculty at Yale since 1956. Dr. Turekian is a fellow of the American Academy of Arts and Sciences, the American Geophysical Union, the American Association for the Advancement of Science, and the Geological Society of America.

NATIONAL RESEARCH COUNCIL STAFF

David A. Feary is a senior program officer with the NRC's Board on Earth Sciences and Resources, and a Research Professor in the School of Earth and Space Exploration and the School of Sustainability at Arizona State University. He earned his Ph.D. at the Australian National University before spending 15 years as a research scientist with the marine program at Geoscience Australia. During this time he participated in numerous research cruises—many as chief or co-chief scientist—and he was co-chief scientist for Ocean Drilling Program Leg 182. His research activities focused on the role of climate as a primary control on carbonate reef formation and on efforts to improve the understanding of cool-water carbonate depositional processes.

Nicholas D. Rogers is a Financial and Research Associate with the NRC Board on Earth Sciences and Resources. He received a B.A. in history, with a focus on the history of science and early American history, from Western Connecticut State University in 2004. He began working for the National Academies in 2006 and has primarily supported the board on a broad array of Earth resources, mapping, and geographical sciences issues.

Courtney R. Gibbs is a Program Associate with the NRC Board on Earth Sciences and Resources. She received her degree in graphic design from the Pittsburgh Technical Institute in 2000 and began working for the National Academies in 2004. Prior to her work with the board, Ms. Gibbs supported the Nuclear and Radiation Studies Board and the former Board on Radiation Effects Research.

Eric J. Edkin is a Senior Program Assistant with the NRC Board on Earth Sciences and Resources. He began working for the National Academies in 2009 and has primarily supported the board on a broad array of Earth resources, geographical sciences, and mapping sciences issues.

Appendix B

Workshop Agenda and Participants

AGENDA

Monday, May 5, 2008

9:00 Welcome Statement — *Dick Norris*

9:30 PLENARY ADDRESS:
Simulating Earth's Climate: Past, Present, & Future — *Jeff Kiehl*

10:30 PLENARY ADDRESS:
Rapid Environmental Change and Feedbacks: Lessons from Deep Time — *Jim Zachos*

11:00-12:30 **BREAKOUT I**

Events and Transitions, Tipping Points, and Thresholds

Question 1: What evidence can we use to identify thresholds and tipping points in the geologic record?

Question 2: What are the best parts of the record to target—and what are the proxies to use—to describe and categorize thresholds and tipping points in the record? What are the nonlinear processes that determine critical "tipping points," and are these processes well represented in climate models and in biota-climate models?

1:30 PLENARY ADDRESS:
Carbon Cycling and Climate Sensitivity Across the Paleocene-Eocene boundary — *Richard Zeebe*

2:00-3:30 **BREAKOUT II**

Coupling and Decoupling Climate Sensitivity

Question 3: What physical and biogeochemical feedback processes are most important in determining the climate sensitivity to a large dynamic range of forcing?

Question 4: What can deep-time records and models tell us about climate sensitivity?

4:00-5:30 **BREAKOUT REPORTS—Questions 1-4**

Tuesday, May 6, 2008

9:00 NSF Hopes and Expectations — *Rich Lane*

9:30 PLENARY ADDRESS:
Dinosaur Forecast—Cloudy! A Convective-Cloud Mechanism for Past Equable Climates and Its Role in Future Greenhouse Scenarios. — *Eli Tziperman*

10:30-12:00 **BREAKOUT III**

Alternative Worlds

Question 5: What are the most poorly understood dynamics of past "alternative worlds," and which "alternative world" intervals offer the greatest potential for understanding future climates?

Question 6: What kinds of proxy evidence do we need to advance understanding of the dominant processes that operate in these "alternative world" intervals?

1:00-2:30 **BREAKOUT IV**

Implementation and Infrastructure

Question 7: Describe the infrastructure that will be required to answer these questions?

Question 8: How do we improve interactions between deep-time data/model research?

Question 9: What are the best options for additional paleoenvironmental and geochronological proxies (e.g., biomarkers and isotopes of biomarkers)?

3:00-4:30 **BREAKOUT REPORTS—Questions 5-9**

4:30 Wrap-up and Thanks *Dick Norris*

5:00 Adjourn

WORKSHOP PARTICIPANTS

Thomas Algeo
Department of Geology
University of Cincinnati

David Beerling
Department of Animal and Plant Sciences
University of Sheffield

Karen Bice
Department of Geology and Geophysics
Woods Hole Oceanographic Institution

Gabe Bowen
Department of Earth & Atmospheric Sciences
Purdue University

Mark A. Chandler
Center for Climate Systems Research
Columbia University
Goddard Institute for Space Studies

Robert DeConto
Department of Geosciences
University of Massachusetts

Harry Dowsett
U.S. Geological Survey

Anthony R. de Souza
Board on Earth Sciences and Resources
National Research Council

David Feary
Board on Earth Sciences and Resources
National Research Council

Alexey Fedorov
Department of Geology and Geophysics
Yale University

Christopher Fielding
University of Nebraska-Lincoln

Margaret Frasier
Department of Geosciences
University of Wisconsin-Milwaukee

Katherine H. Freeman
Department of Geosciences
The Pennsylvania State University

Linda Gundersen
U.S. Geological Survey

Patricia Jellison
U.S. Geological Survey

Kirk R. Johnson
Denver Museum of Nature and Science

Martin J. Kennedy
Department of Earth Sciences
University of California, Riverside

Dennis V. Kent (NAS)
Department of Geological Sciences
Rutgers, The State University of New Jersey

Jeffrey T. Kiehl
National Center for Atmospheric Research

H. Richard Lane
National Science Foundation

Timothy Lyons
Department of Earth Sciences
University of California, Riverside

Isabel P. Montañez
Geology Department
University of California, Davis

Thomas Moore
PaleoTerra

Richard D. Norris
Scripps Institution of Oceanography
University of California, San Diego

Paul Olsen (NAS)
Lamont-Doherty Earth Observatory
Columbia University

Mark Pagani
Department of Geology and Geophysics
Yale University

Martin Perlmutter
Chevron Energy Technology Company

Christopher Poulsen
Department of Geological Sciences
University of Michigan

A. Christina Ravelo
Department of Ocean Sciences
University of California, Santa Cruz

Greg Ravizza
Department of Geology and Geophysics
University of Hawaii at Manoa

Nicholas Rogers
Board on Earth Sciences and Resources
National Research Council

David Rowley
Department of the Geophysical Sciences
University of Chicago

Dana Royer
Department of Earth and Environmental Sciences
Wesleyan University

Nathan Sheldon
Department of Geological Sciences
University of Michigan

Christine Shields
Climate Change Research
National Center for Atmospheric Research

Linda Sohl
Center for Climate Systems Research
Columbia University

Lynn Soreghan
College of Earth and Energy
School of Geology and Geophysics
University of Oklahoma

Christopher Swezey
U.S. Geological Survey

Karl K. Turekian (NAS)
Department of Geology and Geophysics
Yale University

Eli Tziperman
Department of Earth and Planetary Sciences
School of Engineering and Applied Sciences
Harvard University

Thomas Wagner
Institute for Research on Environment and Sustainability
Newcastle University

Debra Willard
U.S. Geological Survey

Scott Wing
Smithsonian Institution

Jim Zachos
Earth and Planetary Sciences Department
University of California, Santa Cruz

Richard Zeebe
Department of Oceanography
University of Hawaii at Manoa

Appendix C

Presentations to Committee

Although most of the community input to the committee occurred at the workshop hosted by the committee in May 2008 (Appendix B), there were several additional presentations by scientists working in the field and by the study sponsors at the first committee meeting in February 2008:

Gaps in Our Knowledge of Past Climates and Causes of Climate Change
Bill Hay
High Frequency Paleoclimate Analysis: Impact on Climate Research and Exploration
Marty Perlmutter
The USGS Perspective on the Committee's Task
Linda Gundersen
Additional USGS Comments
Pat Jellison
The NSF (GEO-EAR) Perspective on the Committee's Task
Ray Bernor
The NSF (OPP) Perspective on the Committee's Task
Scott Borg
Deep-Time Paleoclimate Geoinformatics—A Community Platform for Research and Knowledge Transfer
Walt Snyder
Thoughts on Deep-Time Climate Research
Judy Parrish

Appendix D

Acronyms and Abbreviations

AAAS	American Association for the Advancement of Science
ACEX	Arctic Coring Expedition
CCD	carbonate compensation depth
CIA	Chemical Index of Alteration
CO_2	carbon dioxide
DOE	Department of Energy
DOSECC	Drilling, Observation, and Sampling of the Earth's Continental Crust
DSDP	Deep Sea Drilling Project
EMIC	Earth system models of intermediate complexity
ENSO	El Niño-Southern Oscillation
E&O	education and outreach
ESM	Earth system model
GCM	global climate model; also general circulation model
GIS	geographic information system
ICDP	International Continental Scientific Drilling Program
ICP-MS	inductively coupled plasma mass spectrometry
ID-TIMS	isotope dilution-thermal ionization mass spectrometry
IODP	Integrated Ocean Drilling Program
IPCC	Intergovernmental Panel on Climate Change

ITCZ	Intertropical Convergence Zone
ka	thousands of years ago
ky	thousand years
LGM	Last Glacial Maximum
LPIA	Late Paleozoic Ice Age
Ma	million of years ago
MAT	mean annual temperature
my	million years
μmol	micromole
NAO	North Atlantic Oscillation
NASA	National Aeronautics and Space Administration
NCEAS	National Center for Ecological Analysis and Synthesis
NOAA	National Oceanic and Atmospheric Administration
NRC	National Research Council
NSF	National Science Foundation
OAE	oceanic anoxic event
ODP	Ocean Drilling Program
pCO_2	partial pressure of carbon dioxide
PDO	Pacific Decadal Oscillation
PETM	Paleocene-Eocene Thermal Maximum
PI	principal investigator
PMIP	Paleoclimate Modeling Intercomparison Project
ppmv	parts per million by volume
SIMS	secondary ion mass spectrometry
SST	sea surface temperature
TEX_{86}	tetraether index of 86 carbon atoms; paleothermometer based on the composition of membrane lipids of marine picoplankton
TOC	total organic carbon
USGS	U.S. Geological Survey